DESCARTES' DREAM

DESCARTES' DREAM

The World According to Mathematics

Philip J. Davis
Reuben Hersh

Harcourt Brace Jovanovich, Publishers
San Diego Boston New York

"The Criterion Makers: Mathematics and Social Policy" previously published in *American Scientist*, 50(3):258A–274A, 1962.

"Are We Drowning in Digits?" previously published by UPI wire service, December, 1981.

"Of Time and Mathematics" previously published in *Southern Humanities Review*, 18:193–202, 1984.

"Mathematics and Rhetoric" to be published in *The Rhetoric of the Human Sciences*. John S. Nelson, Allan Megill and Donald N. McCloskey, editors. Madison: University of Wisconsin Press, forthcoming.

The figure on the dust jacket is taken from Michael Maier, *Emblemata Nova de Secretiis Naturae Chymica*, Oppenheim, 1618. It is also reproduced on page 230.

Jacket typography is from *A Constructed Roman Alphabet* by David Lance Goines. © 1982 by David Lance Goines. Reprinted by permission of David R. Godine, Publisher, Boston.

Printed in the United States of America

Library of Congress Cataloging-in-Publication Data

Davis, Philip J., 1923–
Descartes' dream.

Bibliography: p.
Includes index.
1. Mathematics—Philosophy. 2. Mathematics—Social aspects. 3. Computers and civilization.
I. Hersh, Reuben, 1927– . II. Title.
QA8.6.D39 1986 510'.1 86-11967
ISBN 0-15-125260-2

First edition

9 8 7 6 5 4 3 2 1

To Phyllis and to Hadassah
With thanks and love

Contents

René Descartes (1596–1650)
French philosopher and mathematician.

"The long concatenations of simple and easy reasoning which geometricians use in achieving their most difficult demonstrations gave me occasion to imagine that all matters which may enter the human mind were interrelated in the same fashion."
—René Descartes. (*After Franz Hals: Musée du Louvre, Paris. Courtesy of Brown University Library.*)

Giovanni Battista Vico (1668–1744).
Italian philosopher, lawyer, and classicist.

"Mathematics is created in the self–alienation of the human spirit. The spirit cannot discover itself in mathematics. The human spirit lives in human institutions."
—Giovanni Battista Vico. Paraphrase by Sir Isaiah Berlin. (*Classici Italiani, Tomasseo Editorie Tounese, 1930. Courtesy of Brown University Library.*)

Preface

THE GENEROUS RECEPTION of *The Mathematical Experience* has encouraged the authors to continue their description of that wonderfully strange and often baffling activity that is called "doing mathematics." In *The Mathematical Experience,* we tried to give a broad picture of this activity. The point of view was that of the professional, looking at mathematics from the inside and describing the ingredients of which it is constructed, how it is created and evaluated, what it feels like and sounds like to do mathematics, what human values can be assigned to it. In short, we tried to answer the question: what *is* the mathematical experience? In answering this question, we were led to a philosophy of mathematics which we felt was consistent with this experience, and it was the formulation of this philosophy that energized our writing efforts.

The goal of the present book is different. We approach mathematics from the outside. We are concerned with the impact mathematics makes when it is applied to the world that lies outside mathematics itself; when it is used in relation to the world of nature or of human activities. This is sometimes called *applied mathematics.* This activity has now become so extensive that we speak of the "mathematization of the world." We want to know the conditions of civilization that bring it about. We want to know when these applications are effective, when they are ineffective, when beneficial, dangerous, or irrelevant. We want to know how they constrain our lives, how they transform our perception of reality.

Over the past century, mathematics, technology, and business have joined forces in a most marvelous way to produce *the computer*! The computer, recognizing this birthright, has in turn laid many benefits at the feet of its progenitors.

The applications of mathematics are now effected, in their final

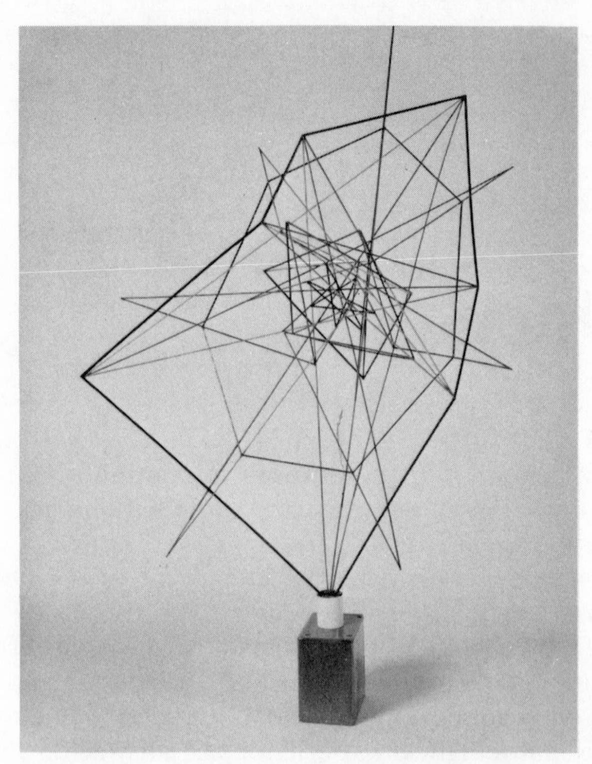

Theorem into Art. I.J. Schoenberg exhibits the harmonic analysis of a skew heptagon (Jesse Douglas' theorem) as a space construction. Can one introduce advanced mathematics as an iconographic element? (*Courtesy of I.J. Schoenberg.*)

stages, through the instrumentality of this marvelous mathematico-logical engine. What is surprising is the wide applicability and acceptance of this technology. The potentialities of the computer are considered to be so all–embracing that many aspects of our day–to–day lives are being reconstituted to fit in with the computer.

We shall ask questions of the computer parallel to those we ask of mathematics. What is the computational experience? How does the computerization of the world affect the physical and intellectual quality of civilization?

As to technology, anyone who surveys the first half century of computerization sees the reciprocal relationship between computers and technology. One doesn't fly to Mars without substantial help from the computer. The public learns that a pacemaker embodies a computer, and it even begins to feel that the computerization of man must be taken in a sense that is literal, not merely metaphoric. Reciprocally, it knows that the bits and pieces, the chips and the tapes, of a computer are products of advanced technology and that every technological advance offers the possibility of improvement in computer design and operation.

When it comes to business, to social statistics, to data processing generally, the public, likewise, sees the mutual play. It sees clearly the revolution that has occurred in bookkeeping. The computer reservation services, though not strictly necessary for the operation of an airline or a theatre, have attained a level of convenience and flexibility we would not willingly forego. The computerization of libraries, which is now in progress, joins all library facilities together and places every book at the fingertips of any scholar who is able to type an inquiry into a terminal keyboard.

When it comes to mathematics, the third member in the partnership, the interaction between it and the computer does not stare us in the face in quite the same way. It is there, all right; in the older view, what is a computer but a machine that does arithmetic rapidly? While business and technology relate to material things, mathematics is a matter of the imagination, a discipline in which abstract symbols are moved around. Abstractions, however, are not immediately comprehensible. Consequently, although the origin of the computer was in the primal dreams of mathematics, and although its subsequent development depended on blueprints supplied by mathematical talent, still the role that mathematics has played is less well understood than the role of technology and business.

To display the relationships between mathematics and computers, we would have to answer many questions. What has been the influence of the computer in bringing abstract mathematical formulations to practical applications? What has been its role in the creation of new mathematics? What has been its influence on theories of mathematical knowledge and existence; on mathematical imagery and insight; on mathematical education? How has it affected our view of what is possible and impossible, and of how the impossible may be transcended?

Reciprocally, how has mathematics contributed to the creation of new computer systems? What are the elements of universality which entitled computer science to be called a science and not merely a craft? Are these elements mathematical? Should theoretical computer scientists adopt as their ideal the deductive program of the mathematician? What is the relationship between mathematical thought, computer potentialities and general human intelligence? In what ways do the dreams of mathematics contrast with those of computer science?

Having answered these questions, we would be well on our way to the creation of a philosophy of computation. This is a subject hardly existing at present apart from borrowings from the philosophy of science or from the philosophy of mathematics. It deserves to exist in its own right.

A philosophy of computation? What could that possibly be? Well, one might say, just to get off the ground, that as classical philosophy was concerned with the true, the good, and the beautiful, so also should the philosophy of computation be concerned with true computation, good computation, and beautiful computation. What makes a computation true? Why should I believe what a computer tells me? What makes a computation useful? What makes it good or bad? What makes it beautiful or ugly? How does the computer change our ideas of reality, of knowledge, of time?

Many reasons may be given for raising these questions. Consider, in the first place, the tremendous momentum which computerization now possesses. We are engulfed by waves of new ideas, equipment, potentialities beyond the dreams of science fiction. In the computer field, although reason cautions us against unlimited optimism, no ceilings are visible; the sky is the limit. There are jobs everywhere, the action is certain, the *esprit* is overwhelming. There is some evidence that the brightest and best of our young people are abandoning the traditional playgrounds of the intellect—mathematics and physics—in favor of computer science. Some of them feel that the mathematical discipline is too rigid and that the subject matter of current research in it is unbearably tedious. In contrast to this, the hell–for–leather mentality of computer science comes as a blessed release. In view of this state of frenetic activity and its accompanying delirium, it would pay us to sit back quietly and ask "What does it all add up to?"

The importance of such questions derives also from social concern. Some critics think that the computer and its ancillary activities are integral parts of the megamachine of contemporary megatechnology and that this machine degrades the spirit and corrupts the intelligence. Simone Weil, a woman of a deeply religious temperament, sister of a great mathematician, wrote in her Notebooks some years before the digital computer:

"Money, mechanization, algebra. The three monsters of contemporary civilization." Money, mechanization, algebra: the three activities we had identified as progenitors of the computer.

Believe, or don't believe, as you will; but her belief is common, and one doesn't have to be an aspiring saint to have arrived at this position. Those of us who love mathematics and revere it as one of the great creations of humanity will want to ask why it is thought that mathematics, in conjunction with its two partners, has given rise to such monstrosities.

At the opposite pole of critical intemperance is the idea that the computer brings the possibility of social salvation through a technological utopia. The true believer asserts that only by the adept and rapid handling of trillions of bits of interlinked information can one hope to arrive at social justice and intercultural tranquility. The ironic believer may assert that by giving man yet another kind of intellectual concern and another toy, by allowing the format of the computer to dominate traditional human concerns, those aspects of belief and action by which men have slaughtered one another to prove the primacy of one idea over another can be phased out to irrelevance.

As one moves towards a philosophy of computation, one should surely be guided by the well–established discussions of the philosophy of mathematics. Over the last century, mathematical philosophy has been concerned largely with one question: *why is mathematics true?* An emerging philosophy of applications and of computation must be concerned with the question: *why is mathematics used*? The movement of our computer culture has been so rapid and so pervasive that it defies the talents of reportage to express adequately where it is, let alone to arrive at the deeper philosophical implications of its existence.

In the present book, we have tried to come to grips, as best we could, sometimes tangentially or trivially, with some of the questions just raised. We had no intention of writing a book whose spirit is "neutral." We want to make a statement and draw a moral.

The statement, in brief, is:

The social and physical worlds are being mathematized at an increasing rate.

The moral is:

We'd better watch it, because too much of it may not be good for us.

Preface

This book is a collection of independent essays grouped loosely around several themes. The essays require different levels of mathematical knowledge, ranging from popular to professional. Readers are encouraged to browse at random and read whatever catches their fancy. Some of the essays are derived from articles already in print, others are modifications of addresses or edited versions of taped interviews.

Acknowledgements

WE SHOULD LIKE to acknowledge the friendly assistance, cooperation, criticism, and encouragement of the following people: Jackie Damrau, Ernest S. Davis, Joseph Davis, Leah Edelstein, Kathy Hall, Shirley Harty, Daniel Hersh, Eva Hersh, Phyllis Hersh, Mark Horton, Jean Jordan, Penelope Katson, Caterina Kiefe, Igor Najfeld, David Pingree, Moira Robertson, Bill Schaab, Patricia Strauss, Peter Wegner, Jerome Weiner.

Particular thanks go to Joan Richards and Charles Strauss for their interest in this book and their willingness to tape a number of interviews for it.

We are grateful to Sophie Freud for discussions over the past year.

We record here with pleasure the generous assistance of the Alfred P. Sloan Foundation and the Wayland Collegium of Brown University in sponsoring a conference on the Philosophy of Computation at Brown University in the Fall of 1982. Thanks go to the following individuals who read papers at this conference: Oliver Aberth, Joseph Ford, Martin Davis, Kenneth Sayre, Dagmar Barnouw, Eugene Charniak, Martin Ringle, Rolf Landauer, Benoit Mandelbrot and Donald Beaver.

The experience gained from this Conference encouraged us to push on with our writing.

Grants from the Alfred P. Sloan Foundation to Philip J. Davis and to Reuben Hersh have been of great assistance in the preparation of this book.

A grant from the Fairchild Corporation has facilitated our studies of computer graphics and art.

We should like to acknowledge the help of the following people in manuscript preparation: Katrina Avery, Ezoura Fonseca, Deborah J. Long, Deborah Van Dam and Roberta Weller. Computer technology may already be phasing out this type of acknowledgement. We worked in the old–fashioned mode, with pen and paper, scissors and paste. The people who helped us worked in a mixed mode with electric typewriters, audio tape, and word processors.

To the typewriter, we owe the staccato style of Hemingway, the typographic humor of e.e. cummings, and the antics of archy, Don Marquis' bug. The next decade will clarify what is gained and what is lost when authors work by themselves and process their thoughts all the way from conception to distribution by computer.

The word processor is one of the most remarkable products of our mathematized age. It may very well change our concept of a book. In the past, an author would write, rewrite, correct, and edit, and finally a book would appear. It might go through several editions. In the word processor era, a book is a dynamic thing which can be continuously updated by the author or his literary heirs. The reader, on line, calls for today's version. A book is converted from a static object to a living institution with all the strengths and weaknesses of such institutions.

We gratefully acknowledge the courtesy of the American Scientist, United Press International, the University of Wisconsin Press and the Southern Humanities Review for allowing us to reprint articles which first appeared in their publications.

Acknowledgement for graphical material is made to The Lownes Collection, John Hay Library, Brown University; John Carter Brown Library, Brown University; Department of Computer Science, University of Utah, Salt Lake City; National Aeronautics and Space Administration; Science 84; Katrina Avery; Louise Goodman; Irene Shwachmann; Wayne Timmerman; Jon Weiner; Frederic Bisshopp; I. J. Schoenberg; Keith Long; Ben Trumbore and Kelan Putbrese.

DESCARTES' DREAM

I

THIS MATHEMATIZED WORLD

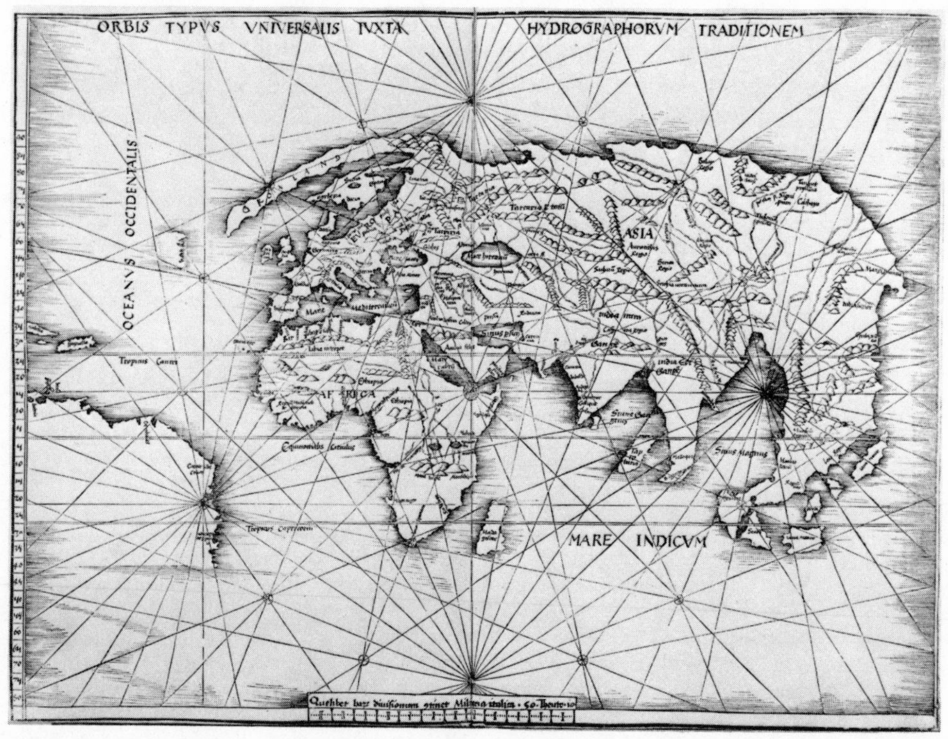

Orbis Typus Universalis. The land mass at lower left is the recently discovered Western Hemisphere not yet designated as America. Here is emergent reality. In: Claudius Ptolemy, "Geographia," Strassburg, 1513. (*Courtesy of the John Carter Brown Library at Brown University.*)

Descartes' Dream

THE MODERN WORLD, our world of triumphant rationality, began on November 10, 1619, with a revelation and a nightmare. On that day, in a room in the small Bavarian village of Ulm, René Descartes, a Frenchman, twenty–three years old, crawled into a wall stove and, when he was well warmed, had a vision. It was not a vision of God, or of the Mother of God, or of celestial chariots, or of the New Jerusalem. It was a vision of the unification of all science.

The vision was preceded by a state of intense concentration and agitation. Descartes' overheated mind caught fire and provided answers to tremendous problems that had been taxing him for weeks. He was possessed by a Genius, and the answers were revealed in a dazzling, unendurable light. Later, in a state of exhaustion, he went to bed and dreamed three dreams that had been predicted by this Genius.

In the first dream he was revolved by a whirlwind and terrified by phantoms. He experienced a constant feeling of falling. He imagined he would be presented with a melon that came from a far–off land. The wind abated and he woke up. His second dream was one of thunderclaps and sparks flying around his room. In the third dream, all was quiet and contemplative. An anthology of poetry lay on the table. He opened it at random and read the verse of Ausonius, *"Quod vitae sectabor iter"* (What path shall I take in life?). A stranger appeared and quoted him the verse *"Est et non"* (Yes and no). Descartes wanted to show him where in the anthology it could be found, but the book disappeared and reappeared. He told the man he would show him a better verse beginning *"Quod vitae sectabor iter."* At this point the man, the book, and the whole dream dissolved.

Descartes was so bewildered by all this that he began to pray. He assumed his dreams had a supernatural origin. He vowed he would put his life under the protection of the Blessed Virgin and go on a pilgrimage from Venice to Notre Dame de Lorette, travelling by foot and wearing the humblest–looking clothes he could find.

What was the idea that Descartes saw in a burning flash? He tells us that his third dream pointed to no less than the unification and the illumination of the whole of science, even the whole of knowledge, by one and the same method: the method of *reason.*

Eighteen years would pass before the world would have the details of the grandiose vision and of the *"mirabilis scientiae fundamenta"*—the foundations of a marvellous science. Such as he was able to give them, they are contained in the celebrated "Discourse on the Method of Properly Guiding the Reason in the Search of Truth in the Sciences." According to Descartes, his "method" should be applied when knowledge is sought in any scientific field. It consists of (a) accepting only what is so clear in one's own mind as to exclude any doubt, (b) splitting large difficulties into smaller ones, (c) arguing from the simple to the complex, and (d) checking, when one is done.

DISCOURS

DE LA METHODE

Pour bien conduire ſa raiſon,& chercher
la verité dans les ſciences.

PLUS

LA DIOPTRIQVE.

LES METEORES.

ET

LA GEOMETRIE.

Qui ſont des eſſais de cete METHODE.

A LEYDE

De l'Imprimerie de IAN MAIRE.

CIↃ IↃ C XXXVII.

Auec Priuilege.

Title page of Descartes' *Discourse.* (*Courtesy of the Lownes Collection, John Hay Library, Brown University.*)

Descartes was first and foremost a geometer; he claimed he was in the habit of turning all problems into geometry. What gives the method substance is the use of mathematics, the science of space and quantity, the simplest and the surest of the conceptions of the mind.

□□
□□

When I was in high school and first heard about analytic geometry —also called coordinate geometry or Cartesian geometry—I thought that it offered a way of reducing any problem in geometry to a corresponding problem in algebra. The problem in algebra would then be solved in a simple and automatic way. The ingenuity required in the classical geometry of Euclid would be eliminated and replaced by an automatic procedure. These ideas must have been in the air, and I simply picked them up. It was even said that analytic geometry was like a huge meat grinder: you stuff the problem in, turn the crank, and out comes the answer.

The truth is not so simple. Anyone familiar with analytic geometry as it is taught today would not recognize it in Descartes' book. What one finds in Descartes is not so much coordinate geometry as the algebraization of ruler–and–compass constructions. Coordinate geometry as currently taught involves the placing of perpendicular axes in a plane, the assignment of two coordinates (or addresses) to each geometrical point, and the replacement of straight lines and curves by appropriate algebraic equations. In its current form, Cartesian geometry is due as much to Descartes' own contemporaries and successors as to himself.

It is true that, in the sense of formal logic, analytic geometry is a machine for deciding the truth of geometric statements automatically. This was established by the logician Alfred Tarski in 1931. Tarski's algorithm is in a format that is not applicable to the geometric questions that come up in practice and is so complex that it would boggle present–day computers. In practice, when one is working analytically, one must often be ingenious in setting up the algebra and in handling its details. Otherwise, the algebra itself can become so formidable as to vitiate the presumed automatic quality of the method.

Nonetheless, for the fierceness and the universality of his vision, and for his philosophy which stressed the role of the thinking individual, it is correct to call Descartes the first modern man and to call ourselves Cartesians.

What was bugging Descartes? Or as the poet Paul Valéry, who stud-

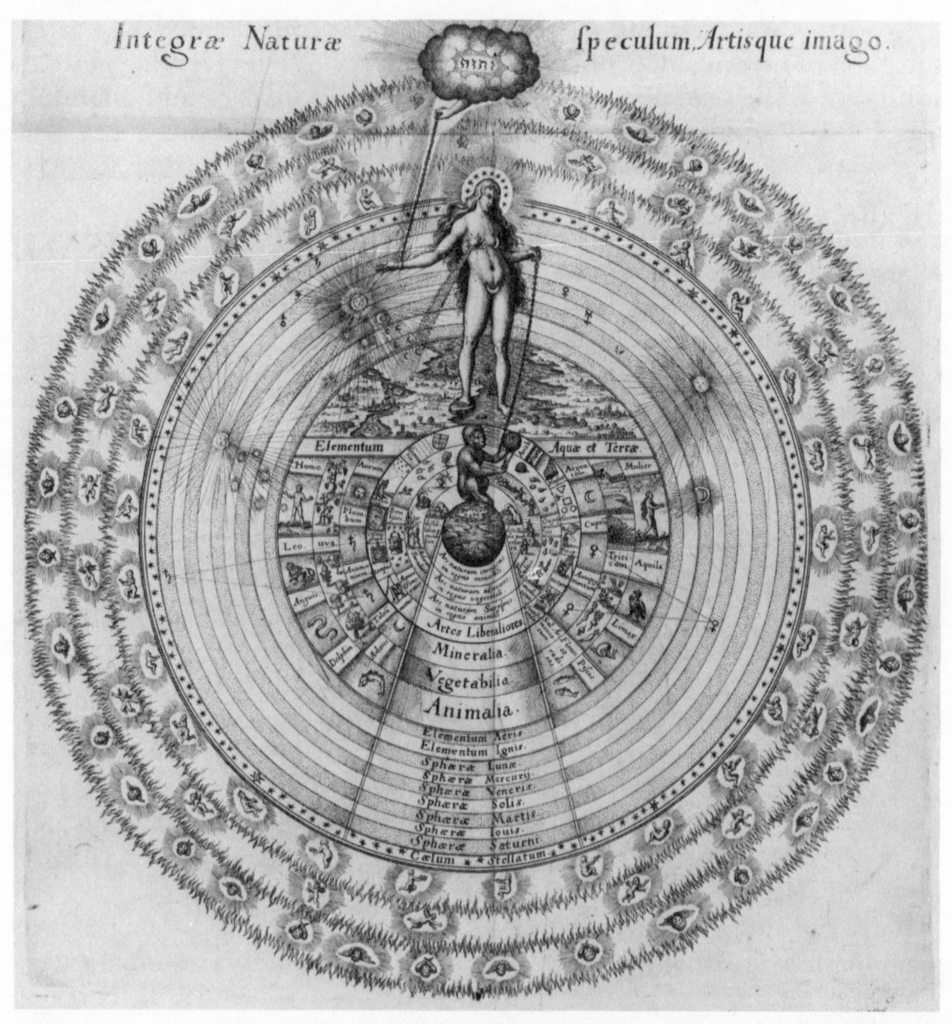

The Cosmos in 1617. Robert Flud's plan. In: "Utriusque Cosmi Metaphysica", Oppenheim, 1617. (*Courtesy of the Lownes Collection, John Hay Library, Brown University.*)

ied Descartes over a lifetime, wrote in his notebooks: "Find what Descartes wanted, what it was possible for him to want, what he coveted, if only half consciously. There's the base, the strategic point to be clarified."

Some scholars think that Descartes had been playing around with the Rosicrucians. Many scientists of that period were members of this brotherhood or at least had studied its doctrines. They think that

Descartes wanted to raise the hermetic and alchemic universalism of the Rosicrucians to the level of precise reason he found in mathematical deduction. More generally, it is thought that Descartes found that the current state of knowledge was an uncritical mixture of fact and fancy, of legend and hearsay, of sense and nonsense, of doctrine and dogma, of experiment, conjecture and prejudice, all infused with stale and ineffective metaphysics and with chaotic and misguided procedures. All this he wanted to reform and revolutionize, to sweep away and replace by a method which was truth–revealing and certain and to underwrite the new science with a new vision and a new philosophy of truth. To this grandiose program he dedicated his life, saying "I have put it above kingdoms and thrones and held riches as naught compared to it."

Without denying this story, I have my own fantasy to explain Descartes' program. I like to imagine that as a young boy he was confronted with a certain problem in mathematics. He tried this and that in order to solve it, and nothing worked. He got stuck. He simply could not solve the problem.

There is nothing peculiar about supposing this. All mathematicians get stuck. The very greatest of mathematical geniuses get stuck. The proof of this statement is evident: there are always famous unsolved problems around. That some of these problems ultimately get solved is irrelevant. In every generation, there is something that the best brains can't do.

Mathematics, said Descartes, is a thing of the mind. Its truths, proceeding, as they do, from sure hypotheses through small, but equally sure, steps of human reason, are guaranteed by God. Why should the mind block itself? If it conceives of a problem it must equally reveal the path along which the solution is to be found.

I like to imagine that having been stuck, a kind of cosmic fury arose in the young Descartes, lasting a lifetime, which he sought to dissipate by finding a method guaranteed always to produce answers. As Valéry observed, "It won him the most brilliant victory ever achieved by a man whose genius was applied to reducing the need for genius."

The vision of Descartes became the new spirit. Two generations later, the mathematician and philosopher Leibnitz talked about the "characteristica universalis." This was the dream of a universal method whereby all human problems, whether of science, law, or politics, could be worked out rationally, systematically, by logical computation.

In our generation, the visions of Descartes and Leibnitz are implemented on every hand.

Cartesianism calls for the primacy of world mathematization.

Further Readings. See Bibliography

W. de la Mare; J. Maritain; P. Valéry; J. Vrooman

Where the Dream Stands Today

LET US TAKE a quick measurement of the march that mathematics has taken in the four centuries since Descartes' dream. To do this, we must have some notion of its extent in Descartes' own day.

In business, the arithmetic of buying and selling had long been in place, as had that of loans and interest. Marine insurance policies have roots in antiquity and were well established by the 15th Century. Casualty and life insurance were coming in strongly in Descartes' lifetime. Lotteries and gambling were an old story, though their deeper theory was just evolving.

In astronomy, the calendar as we know it today was in place, with the exception of several slight corrections. Purely arithmetical methods for calculating the positions of the moon and the planets had been known since antiquity. The work of Ptolemy of Alexandria in the 2nd Century A.D. brought calculational astronomy to a high peak. The geometric schematization of Copernicus and the subsequent studies of Kepler, Tycho Brahe and Galileo would soon lead to the revolutionary work of Sir Isaac Newton wherein, with the development of the calculus, mechanics and planetary motion would be reduced to systems of differential equations.

The measurement of geometric figures, their lengths, surfaces and volumes, had been well known since the days of Euclid (325 B.C.) and Archimedes (225 B.C.). Surveying and certain design problems of architecture were well understood. The geometry of the surface of the sphere and associated problems of geography, mapmaking, and navigation were being fleshed out in the 16th Century. A start had been made in the mathematical theories of optics, perspective, hydrostatics, and hydrodynamics, as well as in the science of musical sounds. Music had been mathematics already to the Pythagoreans, who around 500 B.C. discovered the relationship between pitch and string length. "The harmony of the spheres," an attempt to integrate the experience of music, astronomy and mathematics, was an idea that must have been familiar to Descartes through the recent speculations of the astronomer Kepler. Astrology, which in Descartes' day was allied to medicine, to chemistry (alchemy), and to augury, was highly mathematized. Although it ultimately turned out to be unsuccessful, its methodology of trying to establish relationships was not *a priori* unscientific. It was important in suggesting new problems of applied mathematics and in sharpening computational practices already employed.

The world of Descartes abounded in mathematical instruments. The abacus and the sundial are ancient; their rudimentary forms go back to 3500 B.C. The astronomical quadrant, used to measure angular distances dates from the 800s. The astrolabe, used for the determination of the time of day and the latitude, goes back as far as the 3rd Century B.C. Mechanical clocks date from the 14th Century. At the time of Descartes' dream, John Napier had just published (1614) his logarithms, which advanced the practical art of computation and were ultimately to become one of the basic ideas of theoretical mathematics.

Twelve generations have now passed since Descartes dreamed his dream. How fares his envisioned mathematization of the world? In their theoretical aspects, the natural sciences of physics, astrophysics, and chemistry are now thoroughly mathematical. Indeed, it has become almost a condition of a scientific theory that it be expressible in mathematical language, and it has become almost an act of faith that if the available mathematics is inadequate to describe some observed phenomena the appropriate mathematics can be devised.

The life sciences of biology and medicine are increasingly mathematical. The mechanisms controlling physiological processes, genetics, morphology, population dynamics, epidemiology (the spread of disease), and ecology all have been supplied with mathematical bases.

In sociology and psychology the record is spottier. The accumulation and interpretation of psycho–social statistics is big business, often leading to governmental action. Statistical sampling, polling, and testing may change our commercial and political policies.

Economic theory cannot now be understood without a fair background in mathematics. The theory of competition, of business cycles and equilibria require mathematics of the deepest sort. Game theory, decision theory, optimization strategies may be called on to arrive at commercial and military policy.

It is possible that your retirement fund has made its investments by

(Courtesy of Louise Goodman)

utilizing the newly created portfolio theory and that the quality of our future life on earth will be predicted by the methods of economic time series analysis. Industrial or institutional operations may be laid out by using mathematical scheduling theory.

Linguistics is now more about formal (i.e., mathematical-like) languages than about the compilation of a Navaho–English dictionary. Mathematics has reached into musical composition, choreography, and art.

All computerizations have a mathematical underlay. The digital computer is the mathematical instrument *par excellence*, and a measure of its wide potentialities can be gleaned from this: the IBM Company, whose business it is to make and sell computers, publishes a magazine called "Perspectives in Computing." This magazine contains many articles written by academics and enjoys a wide circulation among them. Its purpose is to point out the ways in which computerization may be introduced profitably in all academic areas, from the analysis and writing of poetry to the collation of Sacred Tibetan texts.

The latest digital recording of Bach's B–minor Mass is produced by filtering acoustic wave forms by means of the Fast Fourier Transform — in chip form. Do you want to understand how a rat learns to tread a maze? Then an appropriate Markoff matrix will tell you, though the rat may complain that its behavior is oversimplified thereby.

Do you want to know how the garbage trucks of New York City should thread their collective way optimally through the streets of Manhattan? Then A. C. Tucker's 1973 paper on perfect graphs will enlighten you.

There have been attempts to give a mathematical definition of life in terms of what is called Complexity Theory. There have been studies which see the tensions between God and man recorded in the Old Testament as instances of game theory. There has been an attempt to put the Problem of Evil into the context of mathematical transform (bypass) theory.

All of this, then, and much much more, is what the Shade of Descartes, returning to earth, would find in the final years of the 20th Century. It should seem, then, that there is hardly an area into which mathematics has not or might not penetrate. Just as all material objects, no matter where they are located, are subject to the law of gravity, just as St. Paul declared he was "all things to all men," so mathematics in its ability to deal with quantity, space, pattern, arrangement, structure, logical implication, has become, as Descartes would have wanted, the unifying glue of a rationalized world.

Further Readings. See Bibliography

C. Boyer; COSRIMS; M. Dertouzos and J. Moses; M. Gaffney and L. Steen; M. Kline; P. Lax; J. Newman; Z. W. Plyshyn; S. Pollack; L. Steen

The Limits of Mathematics

BUT WAIT! Can everything be mathematized? Is there anything in the world which can never become the subject of a mathematical theory? Certainly in the physical world we do not believe there is anything un–mathematizable. There may be phenomena such as turbulence whose mathematical description is so complex we are unable to analyze it or compute it in any reasonably effective sense. We are confident, however, that physics can encompass any physical phenomenon, and do so by means of a mathematical formalism, whether it be the old, familiar one of differential equations with initial and boundary conditions, or the up–to–date one of mappings between high–dimensional or infinite–dimensional non–linear differentiable manifolds.

To find things that cannot be mathematized, then, we must look away from the physical world. What other world is there? If you are a sufficiently fanatical mechanical materialist you may say none. Period. Discussion concluded.

If you are more of a human being, you will be aware that there are such things as emotions, beliefs, attitudes, dreams, intentions, jealousy, envy, yearning, regret, longing, anger, compassion, and many others. These things—the inner world of human life—can never be mathematized.

True, some psychologists and sociologists have come around with their questionnaires and chi–square statistics, purporting to study the human mind quantitatively, but most such investigations are so remote from the target that the critic need hardly say, "Pooh!" They fall over of their own absurdity and pomposity.

I don't mean to say that it is only the inner life of the individual that is beyond mathematics. Even more so is the "inner life" of society, of civilization itself, for example, literature, music, politics, the tides and currents of history, the stuff and nonsense that fill the daily newspaper. All this falls outside the computer, outside any equations or inequalities. And a good thing, too.

Further Readings. See Bibliography

B. Arden; H. Dreyfus (1979); J. Eccles and D. Robinson

Are We Drowning in Digits?

THE POST OFFICE has recently added four digits to its zip numbers. They promise better service, but cannot guarantee it. To call England I must dial fifteen digits (but then I have the thrill of crossing the ocean myself). Institutions installing tricky new phone systems are sending their secretaries to seminars to teach them how to call the office down the hall. For instant money, available twenty–four hours a day, I am encouraged to get a magic card and follow a simple program. I have no doubt that within a few short years, I will have to do some preliminary programming in order to use a public convenience. Putting a nickel in the slot will be listed among the Holy Simplicities of the Past. Are we drowning in digits? Is the end in sight?

Yes, we are, and no, it is not.

What underlies all the digits is that our civilization has been computerized. We are in the grips of the symbol processors and the number crunchers. The nature of this slavery is often misunderstood. It is not thralldom to an individual computer; rather it is the total computerization of the sources of information and communication. Every time a dentist fills a cavity a computer, somewhere, finds out about it and sends a bill. Unplug the computer network? No way. Your son–in–law may have a good job programming the billing system. The dentist himself owns IBM stock.

Numbers and symbol processing; this is mathematics. "Study mathematics! It keeps your options open." Mathematics has joined mechanism and money. Some people think this combination is the monstrosity of the age. Others say: on the contrary, it is the road to salvation. In the New Jerusalem, people speak FORTRAN or BASIC. A computer game can be the new theophany. "I compute, therefore I am" is the new assertion of existence.

We all see the benefits of computers: trips to the moon, pacemakers, intractable mathematical problems solved in a jiffy. We do not yet see the price that will be paid for a state of super–digitalization.

There is occurring today a mathematization of our intellectual and emotional lives. Mathematics is not only applied to the physical sciences where successes have been thrashed out over the centuries but also to economics, sociology, politics, language, law, medicine. These applications are based on the questionable assumption that problems in these areas can be solved by quantification and computation. There is hardly any limit to the kind of things to which we can attach numbers or to the kinds of operations which are said to permit us to interpret these numbers. We are awash in questionnaires, statistics. Standard deviations and correlation coefficients are spat out by computers held in the hands of the uncritical and used as hammers to pulverize us into compliance with the conclusions of the investigator. (Do you think of yourself as deprived? Yes: 17%. No: 48%. Don't understand what deprived means: 12%. Other: 23%.) The Criterion Makers tell us that society should move so that such and such a norm is optimized, and they base policy on this, but no one can say why the criterion is itself appropriate.

Excessive computerization would lead to a life of formal actions devoid of meaning, for the computer lives by precise languages, precise recipes, abstract and general programs wherein the underlying significance of what is done becomes secondary. It fosters a spirit-sapping formalism.

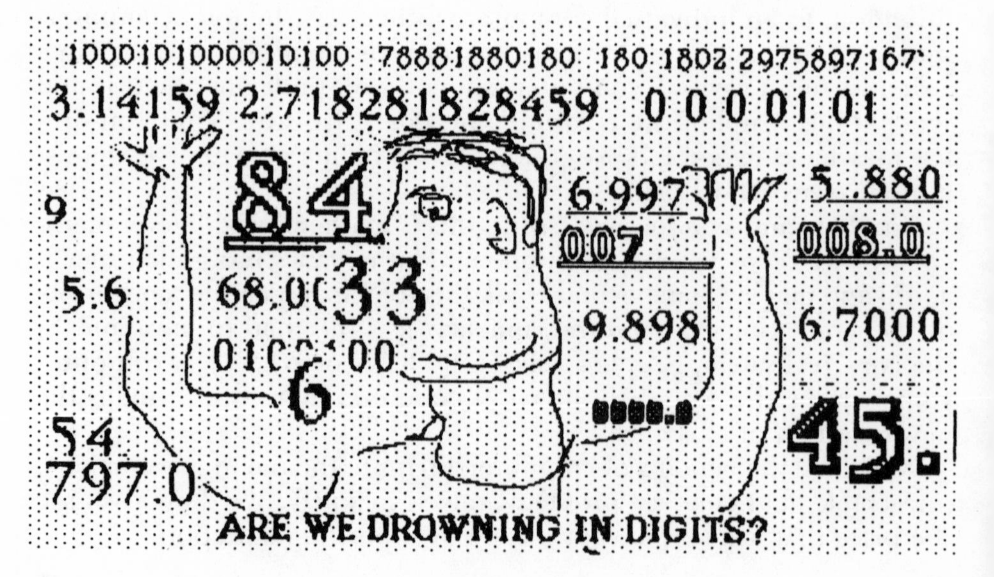

The computer is often described as a neutral but willing slave. The danger is not that the computer is a robot but that humans will become robotized as they adapt to its abstractions and rigidities.

The problem in the coming years is that of establishing meaning in a sea of neutral symbols.

The Stochastized World: A Matter of Style?

A FAIR COIN is tossed, and an action is taken as a result of the toss. The result of the toss is an event of pure indifference. The toss displays complete lack of bias and in its lack of bias is an act at the highest ethical level.

In its indifference, the toss displays complete disengagement from the world of meaning and hence is an act at the lowest ethical level.

In its ambiguity of levels, the toss acquires new meaning.

□□
□□

The word "stochastic," strange sounding but currently very popular in scientific circles, means random, chancy, chaotic. It is pronounced "stoh-kastic." The stochastization of the world (forgive this tongue-twister) means the adoption of a point of view wherein randomness or chance or probability is perceived as a real, objective and fundamental aspect of the world. It refers as well to the utilization of those methods of the theory of mathematical statistics and probability which are intended to reduce the chaos of the single unpredictable event to a less wild and more predictable pattern. The "opposite" of stochastic is deterministic; but we have learned to live simultaneously in a world that is both stochastized and deterministic, so that "complementary" rather than "opposite" describes their relationship better.

One may ask: why use the word "stochastic" when the more common word "statistical" is around? Aren't they identical? The answer is that in today's usage, "statistics" refers to the gathering of quantitative data and the drawing of inferences from it. The word "stochastic" is more comprehensive and refers to the whole conceptual scheme of things, both theoretical and practical, both philosophical and methodological, in which randomness is a dominating feature.

Of the digits that crowd our daily papers many have a stochastic basis. We read about the percentage of families in New York City that are childless, the average number of cars owned by four–person families in Orlando, Florida, the probability that a certain transplant operation will succeed. We read about the odds that Nick the Greek is offering on a certain horserace, a market survey that estimated the monthly gross of a fast–food store in a certain location, and the fact that in a certain insurance pool favorable experience has lowered the rate by $.82 per thousand per month. It is implied that a certain attitude is to be engendered by these disclosures, that a course of action should be set in motion. If it is reported that the English scores of Nebraskan tenth–graders are such and such while that of Iowan tenth–graders are this and that, then presumably someone believes that something ought to be done about it.

The stochastization of the world so permeates our thinking and our behavior that it can be said to be one of the characteristic features of modern life. Our insurance companies, our pension and social security plans, are postulated on notions of randomness. Polling, sampling, election predictions, and scholastic testing are based on stochastic notions, and these are vast enterprises. Mendelian genetics is an extended exercise in mathematical probability theory. "What," asked the talk-show host on the radio, "is the probability that a schizophrenic parent will have a schizophrenic child?" And a doctor who was listening rang up immediately and supplied the figure. Quantum physics is probability, the theory of experimental measurements is probability. The testing of alternatives—is this fertilizer better than that fertilizer, is this medicine better than that medicine—is based on theories of probability and statistics.

A message is being transmitted over microwave: is it received garbled at its destination? Then, how can the probability of transmission error be reduced by the encoding of messages? Mathematicians and communications engineers have written volumes on this one question, and the solution is based on probabilistic definitions of channel capacity plus the most subtle kind of combinatorial analysis.

Theories of pattern recognition are formulated by those who are adept in probability. Epidemiology, the study of the spread of infections, is modeled by stochastic differential equations.

Do you want to know the authorship of the unsigned Federalist Paper? Was it Hamilton, was it Madison? Call up the literary detectives who have run the document through a battery of computerized statistical tests. They will mumble a few words about Bayes' Law and be glad to provide you with the answer.

Decision–making in the face of the uncertainties of life calls for probabilists and gamesmen.

Sports are increasingly mathematical and are being taken over by "stats." Take baseball, for example. One of the leading baseball think tanks has created a numerical index of player worth. How does one arrive at the real value of a player? According to one current scheme, you first calculate a player's offensive rating. This is the number of runs he has created per twenty–five and one–half outs. This figure is then adjusted for the special nature of the player's home park. You then figure a "responsibility" for games based on the number of outs a player has made. Applying the won–lost percentage, you derive a won–lost number for each player. You then combine this with a defensive ranking that takes into account errors, double plays, total chances, and other stats. In this way, you finally arrive at a number that reflects the player's contribution to his team.[1]

When a blue–ribbon national committee was formed to make recommendations in the aftermath of the Three Mile Island atomic disaster, its chairman was John Kemeny, President of Dartmouth and a well–known expert in probability theory.

The world is indeed stochastized and is becoming more so as each day passes. A person can hardly run for Sheriff of Penobscot County without hiring a personal pollster. The coloration of a new toothpaste, whether it should be zebra–striped or polka–dotted, may be the object of a vast market survey, for millions of dollars may be put at risk.

The stochastic view of the world may be built into law. In certain states, the selection of juries has a stochastic element. If we wish to register our car, we must buy car insurance, if we wish to obtain a mortgage on a home, we must buy homeowner's insurance, if we wish to run a business of a certain type we must buy casualty insurance; and, of course, the existence of the insurance policy, whether mandatory or elective, feeds back into the system. Patients become litigious, lawyers fan the sparks of greed, doctors take out malpractice insurance. Juries, aware of this, award incredible damages which raise the insurance premiums to unheard–of levels, which raise costs all around and contribute to economic instability.

The stochastic spirit says, "Are there auto accidents? Well, tough. Cover yourself from head to toe with insurance." The deterministic spirit would say, "Analyze the causes. Make changes. Pass laws. Get the drunks off the road." Prudence says: do both.

The stochastic spirit permeates the business world in another way. Is capital based upon swashbuckling economic adventures, or is it really a mode of spreading risk? Many economists think the latter, and they point to the formation of conglomerates wherein the high profits in one large business are balanced against the low profits or losses in another.

Yet the strange thing is that this view of the universe, for all that it is now so pervasive, is relatively new. It is hardly four hundred years old. Probability theory and statistics don't really get off the ground until we are willing to perform one of the most elementary of mathematical operations and accept the consequences of that act. That operation is the process of forming an average. Say there are 5 men in a room. Their weights are, respectively, 161, 173, 154, 192, 168 pounds. Their average weight is 169.9 pounds. Their average weight is the weight of no individual in the room. We create a fictitious man: "the average man." There is no such person. Three men are below average, two above. Not everyone, despite what the congressman asserted for his district, can be above average. We create a new logic and a new epistemology wherein "the reasonable man" is interpreted to mean "the average man."

A fair coin is tossed three times. What is the probability of getting precisely two heads? The answer is 3/8, and this value may be thought of as the averaging of the number of favorable cases over the total number of possible cases.

An individual event may be unpredictable, but when many events are averaged one arrives at stability, order and lawful behavior. It is incredible that the process of averaging was not common until the seventeenth century. The delay in the arrival of the theory of probability is one of the enigmas of the history of science. It is significant also that the birthdate of probability coincides pretty much with the birthdate of the comprehensive mathematical theories that put forward the other great world view—the deterministic mechanics of Galileo and Newton. These two alternatives have always advanced together in a strange process of mutual support and mutual rivalry.

It is worthwhile to review the reasons that have been advanced for this delay in the arrival of probability. Here I shall follow the summary given by Ian Hacking in his witty and perceptive book *The Emergence of Probability*.

There have been, according to Hacking, five main explanations of why probability theory arrived so late on the scene. According to the first view, there was an obsession with determinism and personal fatalism. Once the wheels of the cosmos had been set in motion, everything subsequent was determined and hence potentially knowable. But how was the future to be known? A common method over the centuries has been randomization. Throw the dice, read the cards, cast lots, look at the flights of birds, at the entrails of a sacrifice; in short, use any of the algorithms of diviners, necromancers, seers.[2] The consultation of a random element was often thought to reveal the will of God (do we not today toss a coin to insure ethical fairness?).

According to the second view, it was believed that God speaks through randomization, hence to construct a theory of the random would have been to commit an impious act. In the ancient world, where life and religion were inseparable, such a theory could not have thrived.

The third explanation goes like this: in order to formulate a theory of probability, we need to have available numerous easily understood empirical examples. We need many instances of equiprobable sets of events. In the ancient world, where even the dice were made of irreg-

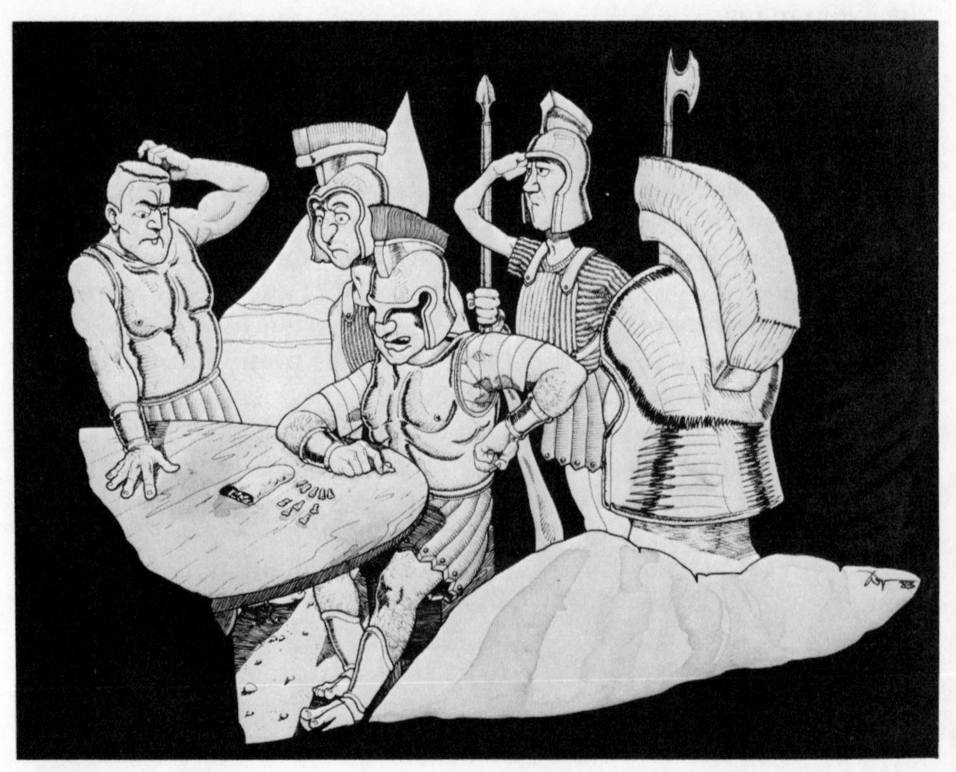

The prehistory of probability theory: Roman soldiers playing bones. (*Artist: W. Timmerman.*)

ular animal bones and not standardized cubes, sets of equiprobable events were not noticed.

According to the fourth view (Marxist), science develops according to economic needs. Not until the seventeenth century do we have the beginning of annuities and insurance. Only in the eighteenth century are there developed theories of measurement, and not until the nineteenth century do we have biological experimentation and data as well as statistical mechanics. All of these things fed into the means and processes of productions of society and established a social need for the theory of probability and statistics; earlier than the seventeenth century there was no such need.

Finally, in the fifth view, mathematics before the Sixteenth Century was not sufficiently developed to allow applications to probability. Techniques of arithmetic were still primitive, and calculus, which is necessary for the description of probability distributions, was not available till the middle of the seventeenth century.

All of these explanations are, in Hacking's opinion and in the opinion of most scientific historians, insufficient to explain the mystery of delay. The jury, composed of historians of science, listens to all the arguments and brings in the verdict: not proven. Now, a new explanation is emerging based on the nature of what is considered to be 'knowledge'. Sometime in the sixteenth and seventeenth centuries, that most fertile period in the history of western thought, a subtle and ultimately profound shift occurred in the nature of what was considered to be evidence, signs, opinion, knowledge, belief.

Hacking himself tends to a theory which says that a new concept of scientific evidence came in through the experiences in the so–called "low" sciences: alchemy, geology, medicine. Authority was relocated in nature and not in the mere words of Authorities, and this led to the practice of observing frequencies.

Lorraine Daston, in a recent study, elaborates the role that Roman–canonical jurisprudence played in the formative days of probability theory in the 17th Century. With its theories of evidence, with a hierarchy of credibility for testimony and proof, with a concern for aleatory contracts (those of insurance, annuities, games of chance, and commodity futures), with a concern both for equity in exchange and for the traditional laws against charging interest, law was uniquely placed to make an impact on science through stochastization.

I should like to set forth my own understanding of how law enters into probability. I shall start by describing *Luca's Problem.*

This is one of the first recorded problems in the history of the theory of probability. It is found in the book *Summa de arithmetica* . . . of Fra Luca Pacioli, Venice, 1494. Pacioli was a mathematician and a friend of Leonardo da Vinci. I have modernized the language a bit.

> *A* and *B* are tossing pennies. *A* wins a toss if it comes up heads. *B* wins a toss if it comes up tails. A stake of $100 is put up, and an agreement is made that the first person to win six tosses wins the whole stake.
>
> Due to an outside interruption, the game has to be broken off after *A* has won five tosses and *B* has won three tosses.
>
> Question: How should the stake be divided?
>
> Pacioli argued that 5/8 of the stake should go to *A* and 3/8 to *B*.
>
> Other authorities, contemporaries of Luca, disagreed.

From the vantage point of today's highly developed theory, one can, I think, distinguish fairly clearly three aspects of probability theory. They are

1. Pure probability theory; this is mathematical, axiomatic, deductive. The statements here have the same epistemological status as in any branch of pure mathematics.
2. Applied probability theory; this attempts to fit probabilistic models to real–world situations. It combines experimentation and data collection with 1) to arrive at statements about the real world.
3. Applied probability at the bottom line. Here the idea is to make practical decisions. What shall we believe in consequence of 1) and 2)? What actions should we take as a result of 1) and 2)?

3) isn't really mathematics at all. It is decision, policy–making, public or private, backed up by the mathematics of 1) and 2).

Textbooks in probability theory are good on points 1) and 2). They are notoriously poor on 3). The passage from 1) and 2) to 3) is accomplished by art, cunning, experience, persuasion, misrepresentation, common sense, and a whole host of rhetorical, but non–mathematical, devices.

A passage from 1) and 2) to 3) is by no means always clear or compelling. "Between the idea/And the reality/Between the motion/And the act/falls the Shadow."

Witness: Gambling. The message of probability theory about gambling is disappointing. It says: mathematics can define something called a fair game. If the game you play is an unfair game, get on the right side of it. If the game is fair, don't bother to play. Mathematics (apart from applied statistics) has very little advice on whether a specific real world game played on specific dice or wheel is or isn't a fair game.

Witness: Smoking.

Witness: People build their houses on the slopes of Mt. Etna or on the San Andreas fault. When a calamity results, a national disaster area is declared and low interest loans, paid for by more prudent citizens, are supplied.

Luca's problem comes before there are any clear views as to what mathematical probability is. It is interesting precisely because it is phrased in terms of 3): what action should be taken as a result of a certain probabilistic occurrence?

It reveals an emerging feeling that there must be a "right way," a "fair" way, an "honest" way, a mathematically deducible way of arriving at a proper division of the stake. This fits in with the opinion that demonstrative knowledge is possible, a feeling that culminates in the idea of Leibnitz that deductive, computational ethics are possible. Today's legal profession might laugh at this.[3]

Luca says, divide the stake in the ratio 5/8 to 3/8 or .625 to .375. Today's probabilist would answer differently. Here is his line of thought. He imagines that the game is played forward to a conclusion. One then has the following "possibility tree": At each stage there are two possible outcomes, heads or tails, each with probability 1/2 or .5 (because the coin is assumed to be fair). In each box we have written next to A the number of heads that have come up so far, and next to B the number of tails. We start with the state of play when the game was interrupted ($A = 5$, $B = 3$) and continue until either A wins with 6 heads or B wins with 6 tails.

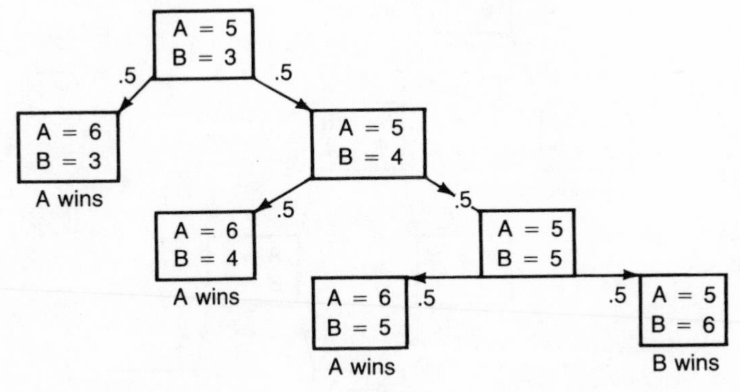

The diagram shows that there are three ways for A to win. Adding up the probabilities of these three distinct outcomes we get $.5 + (.5)^2 +$

$(.5)^3 = 7/8$, while the probability of *B* pushing through to a win is $(.5)^3 = 1/8$. Therefore, the stake ought to be divided in the ratio 7/8 to 1/8 in favor of *A*.

Let us spell this out a bit more. One imagines an infinite sequence of games played through to completion. Now examine those games in which, at some stage, *A* has won five tosses and *B* three. Then ask: of those games, what fraction did *A* ultimately win? The mathematical answer is seven–eighths of them. How do we pass from this mathematical computation to the (judicial) recommendation of a 7:1 split? This depends wholly on the acceptance of the above imaginary process as leading to an equitable division. What is fair is what happens on average. The question is then referred back to what the "experience" is to be averaged.

For example, insurance rates are now calculated separately for men and women; this distinction is one that is made by choice, not compulsion. One could equally calculate rates for the population without regard to sex as is now done, for example, without regard to ethnic origin. The rates would come out differently. The choice of what to average over is not part of the mathematics; it is made before the calculation begins.

Let us return to Luca's problem. On the basis of 1) and 2), one might argue to a different answer. We can equally well suppose we do not know in advance whether the coin is fair or not. This is something mathematics cannot tell us. Assume that the probability of tossing a head is p and tails is q. The tree looks like this:

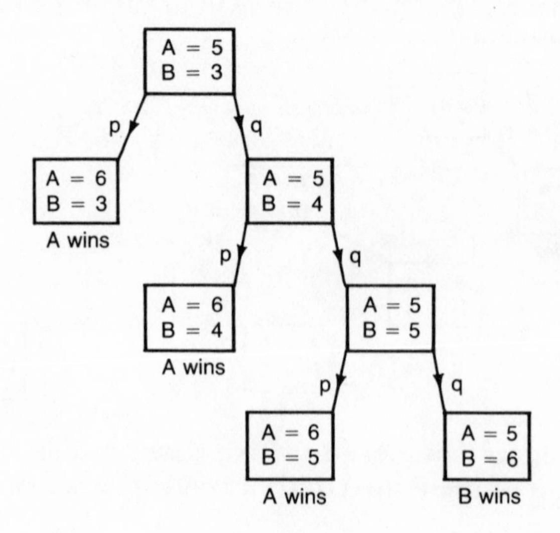

Then the probability of *A* winning is $p + qp + q^2p = 1 - q^3$ and the probability of *B* winning is q^3, so that the division of the stake should be $1 - q^3$ to q^3. So far we don't have a numerical answer because q, the probability of tails, is still unknown.

Now what is the best that one knows about p and q? Only that in the eight tosses that actually took place, there were five heads and three tails. Obviously, one should take $p = 5/8$, $q = 3/8$. Calculating $1 - q^3$ to q^3 with $q = 3/8$, we find that the split of the stake should be .947 to .053.

Are these two answers 7/8 to 1/8 and .947 to .053 the only possible ones? Certainly not. One can argue that the split should be 1/2 to 1/2, on the basis that the original contract is invalidated by the incompleteness of the game. A good lawyer might even make out a case for giving all to *A*. After all, the intent of the game was winner take all. Now who should take all? Certainly not *B*. I suspect that one could write down postulates which would justify any split whatsoever.

Why was the game terminated? The hour was late? A fire broke out in the casino? One of the players died? Was it terminated by *B* because he suspected that *A* was cheating? A lawyer would want to know. The answer matters.[4]

I should like to suggest that progress in probability may have been delayed by just the type of thinking required by Luca's problem. It goes immediately to bottom line probability 3) and raises a question that is not uniquely answerable. Probability theory could only make progress when it was found possible to lift it out of the arena of public policy and experience to develop an idealized and conventionalized version wherein unique answers universally agreed upon are possible.

The probability theory we know in today's mathematics began when inveterate gamblers of the fifteenth century sought help in rationalizing their experience. Despite the fact that we live today in a stochastized world, the underlying meaning of this stochastization has remained elusive and controversial. The mathematics is firm enough in the formal, deductive sense, but to just what aspect of the real world does it correspond?

Is probability real or is it just a cover–up for ignorance? The question of what is real is seldom easy. Is the Devil a real aspect of the world? In centuries gone by, the answer was clearly yes. Today, in the developed world at least, the Devil has receded to a more modest and metaphorical role. Chance is often called into account for the sort of thing that was once attributed to the Devil's dirty work.

From an old American arithmetic book.

There were birth pangs. There were controversial definitions that had to be hacked out and accepted. Take the notion of probabilistic expectation, for instance.

> "Throughout the 18th century, the material of probabilistic expectation was a matter of controversy among mathematicians. Despite its seminal role in the earliest formulations . . . , expectation did not remain a fixed concept, but underwent several striking changes in definition. (It was) altered by mathematicians in a deliberate effort to capture the salient concept of rational decision making. As the notion of rationality successively took on legal, economic and then psychological overtones, the definition of probabilistic expectation followed suit."—L.J. Daston

In a famous quotation rejecting the stochastic aspects of quantum theory, Einstein said, "God does not play dice with the world." In addition to rejecting the metaphysics of quantum theory, this pronouncement also seems to contain an ethical judgement, that the affairs of the cosmos are not arranged in the manner of the shabby entertainments of mankind. When, as a student, I first heard this quotation, it was in an anti–Einsteinian context, with an implication that Einstein's old–fashioned determinism had seen its day.

I should like to mimic the words of R.H. Tawney in his introduction to Max Weber's *The Protestant Ethic and the Spirit of Capitalism.* In this quotation, I have substituted the word "stochasticism" for "capitalism" and made a few excisions.

> "All revolutions are declared to be natural and inevitable once they are successful, and stochasticism is clothed today with the unquestioned respectability of the triumphant fact. But in its youth, it was a pretender, and it was only after centuries that its title was established. For it involved a system of relations that were sharply at variance with venerable conventions. So questionable an innovation demanded of the pioneer who first experimented with it as much originality, self-confidence, and tenacity of purpose as is required today of those who would break from the net that it has woven."

I hear rumblings now of a turnaround. The following anecdote was told recently. One of today's leading theoretical physicists (name supplied on demand) was meeting with some of his colleagues. He said, "You want to know something?" The colleagues nodded. He looked around and said, "Close all the doors." When the doors were closed, he said, "You know, I don't really believe quantum mechanics at bottom."

□□
□□

The most medieval man that I ever met — with respect to the stochastic world — was my cousin H. Having made a lot of money in his younger days as a manufacturer, Cousin H., a bachelor, a chemist, and a cracker–barrel philosopher, spent forty happy years at the racetracks of New England. Statistics, probability were to him of little account. The toteboard odds at the parimutuel tracks were the reflection only of the prejudices of the betters plus the placements of large amounts of cash by the "layoff" men. To place bets according to the wisdom of the toteboard was to invite boredom, was to misread the universe, and was ultimately to misspend one's life.

"Parimutuel has been the ruination of clean racing in the USA," he would tell me. "In the old days, the individual bookie had his livelihood on the line and that tended to keep the races honest. Now the bookmaker at the track has disappeared and the races are as crooked as a pretzel. It adds a layer of difficulty that I have little need for in my declining years, but I try not to allow it to interfere with my enjoyment of the track." Then he would reminisce about the "good old days" at Saratoga where one sat on the long piazza of the U.S. Hotel sipping planter's punches and swapping information with the passers–by.

He would harp on other tunes. "Your probability theory is a bust." (He was a graduate of MIT) "It doesn't tell me what I want to know. It tells me if $100,000 is bet on one race and the track gets 10% and the state gets 10%, $80,000 goes back into the pockets. Now, compound that over eight races, and you'll stay away from the track. From the track's point of view, there is no probability at all. If $10,000 is bet on Sweet Rosie at the rate of $8 to $1, the track distributes $80,000 to her backers. Where's the probability?" Or he would get off racing and onto Wall Street. "If you read about it in the paper, it's too late." In other words, once the average man has got his paws on something, forget it. He disdained the vast quantities of statistics printed in the pages of the *Daily Racing News* (but he always bought a copy as the hallmark of racetrack respectability). He disdained Clocker Walker's Selections and Clocker Tocker's Tips, tout sheets printed on tiny presses located in the backs of station wagons. He disdained the computerized predictions that crunch the stats into recommendations.

What Cousin H. sought was private knowledge, not probabilities. One horse was going to win the race; that much was certain. Which one? In order to answer this question, he set up his own informal network of knowledge. At each of ten New England tracks he had many friends among the owners of horses, grooms, jockeys, stablemen, trainers, handicappers, barbershop assistants, hotdog sellers, and it was there that he sought the answer to the big question. While other patrons would be busy wetting the tips of their pencils and figuring the race on the basis of Lord knows what, Cousin H. would be patiently making the rounds of the bars and paddocks, slowly assessing what he had heard. He did not bet on many races—perhaps one or two a day—but when he decided that his information was overwhelming, he would bet big (big from my point of view; where I would bet $2.00, he would bet $200.00).

Over the years, his track record was, in his own words, "just good enough to keep me entertained."

Every Sherlock Holmes has an Irene Adler as his nemesis. So also Cousin H. Let me call her simply Irene Adler. She was a professor at Simmons College. Irene, also unmarried, had somewhere along the line picked up a taste for the ponies. The two met frequently at the track. She worked with the racing forms and the handicapping sheets and, of course, had her own system. "Her system is absolutely stupid," Cousin H. reported, "but she makes money consistently. She doubles her salary. The IRS knocks at the door but can't find a way in. *She* calls what she's doing rational, but don't believe it. She's been led on by Macbeth's Witches and her comeuppance is just a short way down the road."

"Why don't you go along with her selections? She doesn't seem to be secretive about them."

"You don't understand. It's a matter of style."

This tale is an attempt to come to an understanding of a de–stochastized way of life, a way that says: I will risk all. I may gain all, I may lose all, but I won't join the crowd and average out. The stochastic view has so engulfed us that we would feel absolutely unprotected and naked to the world if we were compelled to come out from behind our averages. Probability is a net that supports us and a cage that confines us. Without it the scientific model makers would be compelled to go back and try again to coax the patterns of the macro–world out of the deterministic dances of the micro–world, an effort which has thus far turned black hair gray.

Notes

1. Sportswriter Roger Angell rhapsodizes on the use of 'stats' in baseball: "The box score, being modestly arcane, is a matter of intense indifference, if not irritation, to the non-fan. To the baseball-bitten, it is not only informative, pictorial, and gossipy but lovely in aesthetic structure. It represents happenstance and physical flight exactly translated into figures and history. Its totals—batters' credit vs. pitchers' debit—balance as exactly as those in an accountant's ledger. And a box score is more than a capsule archive. It is a precisely etched miniature of the sport itself, for baseball, in spite of its grassy spaciousness and apparent unpredictability, is the most intensely and satisfyingly mathematical of all our outdoor sports. Every player in every game is subjected to a cold and ceaseless accounting; no ball is thrown and no base is gained without an instant responding judgement—ball or strike, hit or error, yea or nay—and an ensuing statistic. This encompassing neatness permits the baseball fan, aided by experience and memory, to extract from a box score the same joy, the same hallucinatory reality, that prickles the scalp of a musician when he glances at a page of his score of *Don Giovanni* and actually hears bassos and sopranos, woodwinds and violins."—"The Summer Game", Viking Press, N.Y., 1972.
2. A more drastic method available to the ancient world was the use of magic. The

objective here is to compel the universe to deliver up certain results. How does one compel the universe? Today, through physical means, a river is dammed to yield its energy, a living body is drugged to perform at a higher level. In the fourth century angels could be employed as intermediaries. There is a recipe for fixing a horserace that is so juicy that I cannot forbear quoting it. It occurs in the Sefer ha Razin, a Jewish text on magic of approximately the fourth century:

"These are the angels that serve Rahihel: Agra, Zargir, Genetos, Ta'azama, Zetesrafael, Gadiel, Tammiel, Akahiel, Guchpaniel, Arkani, Zapikuel, Mushiel, Susiel, Harniel, Zachriel, Achnaset, Zadkiel, Achset, Nichmara, Padriel, Kaliliel, Dromiel.

"If you want to run horses fast, so that they won't lose their races, so that they will be fast as the wind, and no animal will put foot before them, and they will be graced by God in their race: take a piece of silver and write on it the names of the horses and the names of the angels, and to the angels that are set over them say: 'I adjure you, oh Racing Angel, who runs between the stars, to strengthen the horses in such and such a race and the charioteer so and so who is running them, so that they will run without fatigue and without stumbling and they will run and be as fleet as the eagle and no animal will stand before them and no magic or witchcraft will prevail against them.' And take the piece of silver and hide it in the race which you want to win."

This of course is anti–stochastic.

3. See Note #4.
4. Cf. Michael O. Finkelstein "Quantitative Methods in Law", New York: Free Press–Macmillan, 1978, for a remarkable set of studies of the applications of mathematical probability and statistics to legal problems. See also the Appendix for a debate over the use of mathematics in the law of evidence. Finkelstein, who is a member of the New York Bar and has collaborated with the eminent statistician Herbert Robbins, summed up his experience in a letter to the author as follows: "Over the years I have come to feel that statistics in law is not as satisfying a subject as I first thought it was; it leads to no firm conclusions, only more and difficult questions. My consolation is that possibly the law is better off with statistics than without it, although I am no longer as sure of this as my writings would indicate."

Further Readings. See Bibliography

C. Bennet; G. Chaitin (1975); L. Daston (1979, 1980); F. David; S. Eitzen; T. Fine; M. Finkelstein; J. Ford; C. Gillispie (1963); I. Hacking; W. Lineberry, D. Owen; S. Siwoff; M. Townend; M. Weber

Feedback and Control: The Equilibrium Machine

WITH A LITTLE BIT of effort one can easily put together the "equilibrium machine" illustrated in the accompanying figure. All one needs from a chemical supply shop are a few bottles, tubes, petcocks, stands, and a variable–speed pump. Water is pumped from the reservoir at the bottom into the top bottle. From there the water flows under gravity through a sequence of subsidiary bottles and ultimately makes its way back to the reservoir. It may look bewildering, but in principle it is no more complicated than the Fountain of Trevi or a recirculating water system for a goldfish tank. Yet this apparatus can serve to demonstrate in a striking way many important concepts: closed system, feedback, conservation, automatic control, equilibrium, steady state, stability, optimal control, catastrophe, difference equation, differential equation, and mathematical modeling in all its difficulties.

The hydraulic mechanism is in open view, and so is more easily grasped by the scientifically inexperienced than a corresponding electronic system. The apparatus is interesting to play with and stands as a vivid metaphor or a model for all kinds of other systems that have nothing to do with bottles: the economy*, a chemical plant, circulation of blood in the body, an organization that hires employees when they are young and retires them at age 65, ecology, cosmology, etc.

Focus your attention on the circulating water. Suppose it is always confined within the tubes and bottles: this is a closed system. If a bottle inadvertently overflows, a catastrophe results, and the floor gets wet. It is arbitrary what is considered to be the extent of the closed system: if one chooses to include the floor, the resulting theory will be different.

* In the late 1950s, A.W. Phillips of the London School of Economics made such a machine to illustrate a Keynesian model of equilibrium.

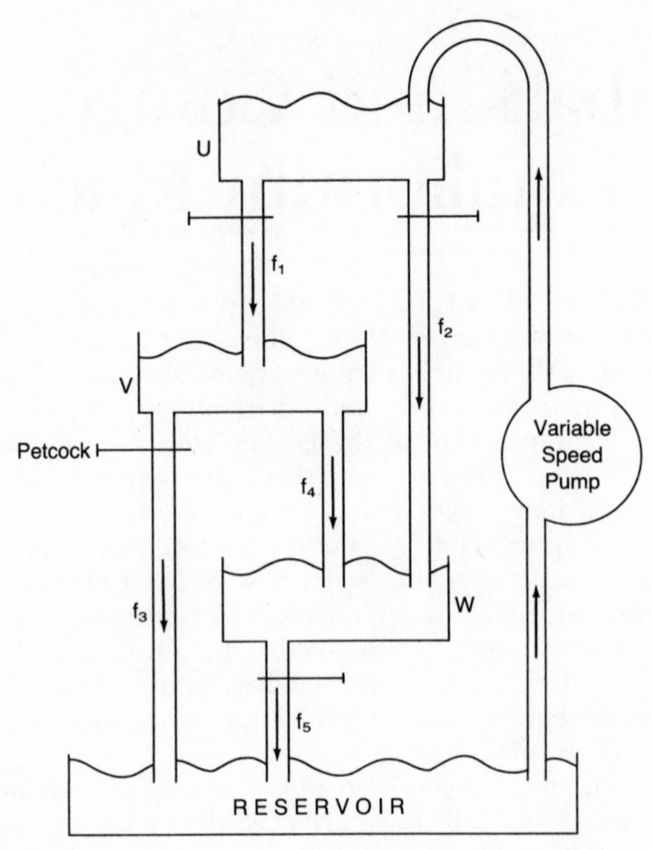

Experiment to illustrate feedback, equilibrium, steady state, stability, control (or lack of it).

The Long Run

Our apparatus is in place. The petcocks are opened to a certain level. These are the controls. We turn on the pump at a certain speed. The water quickly rises in the delivery tube and falls through the discharge tubes. The water level rises in the upper vessels and sinks slightly in the reservoir. What happens in the long run? It is important to know. The custodian of the Fountain of Trevi does not want to sit watching it all day long just to avoid catastrophes.

We know from experience that if our adjustments are good, the machine may tend to a state of equilibrium, after which no further

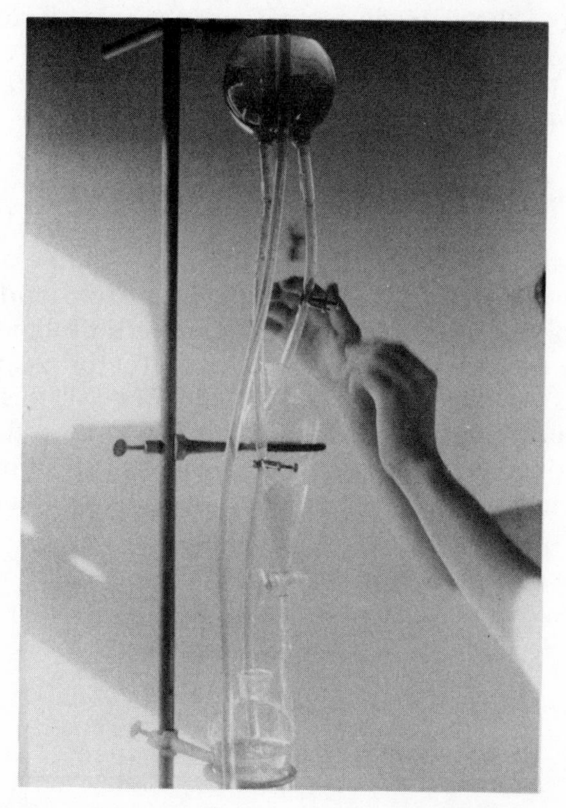

Trying to arrive at equilibrium. (*Photo courtesy of Jon Weiner.*)

change in the water levels will be seen. When the machine is in equilibrium, the casual observer, looking at it from a distance, might be deceived into thinking that nothing is happening. On the other hand, the water level might rise in one of the containers until the water spills over onto the floor. Water is then lost from the reservoir. The machine may then be able to equilibrate itself with a smaller amount of water.

Torricelli's Law

According to theoretical hydraulics (Torricelli's law), if a vessel has a tiny hole at its bottom, then the water will emerge from the hole with the velocity $\sqrt{2gh}$, where g is the gravitational constant 32 ft/sec^2 and

h is the height of the water in feet. Hence the rate of outflow in cubic feet per second is this exit velocity multiplied by A, the cross-sectional area of the hole: rate of outflow $= A\sqrt{2gh} = A\sqrt{2g}\,\sqrt{h} = \text{constant} \cdot \sqrt{h}$. To make this theoretical formula conform better to an experiment, it has to be multiplied by an empirically determined constant which takes into account the shape of the hole. For a vessel of fixed shape and size, the volume of water in the vessel is determined by the height of the water, and vice versa; if you tell me the volume of water, I can tell you, by using geometry, how high the vessel stands. If we let v denote the volume, we obtain a formula of the following type: the outflow in one short interval of time equals $f(v)$ for a certain function f characteristic of the vessel and the hole. We call this the "outflow function." The interval of time should be taken so short that the volume v of water in the vessel does not change very much. When $v = 0$, the outflow, of course, is 0. A typical behavior is illustrated by the figure.

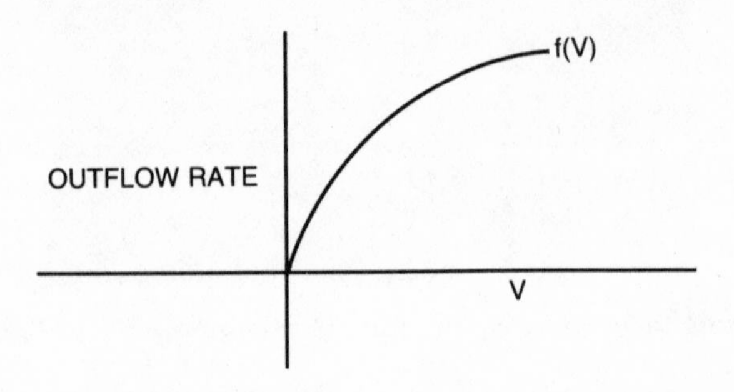

As the volume v increases, that is, as we move to the right along the horizontal axis, the height of the graph $f(v)$ increases. This is because the larger the volume, the higher the liquid stands, and hence, by Torricelli's Law, the more rapid the outflow. The outflow rate is therefore an increasing function of the volume. The total outflow rate from two or more orifices also has this characteristic.

Referring back, we assume that the amount of water in each bottle is observed at discrete instances of time. The bottles are designated by U, V, W and the volumes of fluid present in the nth time interval are designated by U_n, V_n, W_n. The outflow functions for the first bottle are designated by f_1 and f_2 because U has two outlets. For V they are f_3 and f_4, while for W the function is f_5. R_n will be the amount of water in the reservoir at time n and P_n the amount delivered by the pump from the reservoir to the bottle U between time n and time $n + 1$.

The Conservation Laws

Fixing our attention, for example, on the highest bottle, we can see that the volume of water in the bottle at observation time $n + 1$ equals the volume that was there at time n, plus the amount delivered to it between time n and time $n + 1$, minus the amount drained out in that time interval.

This statement can be expressed algebraically as an equation:

$$(*) \qquad U_{n+1} = U_n + P_n - f_1(U_n) - f_2(U_n)$$

Three more equations, similar in form, will hold for the other bottles V and W and the reservoir R:

$$(*) \qquad \begin{aligned} V_{n+1} &= V_n + f_1(U_n) - f_3(V_n) - f_4(V_n) \\ W_{n+1} &= W_n + f_2(U_n) + f_4(V_n) - f_5(W_n) \\ R_{n+1} &= R_n + f_3(V_n) + f_5(W_n) - P_n. \end{aligned}$$

These equations will be true for each time interval $n = 1, 2, 3, \ldots$. They are called the "continuity equations" of our machine. This term is borrowed from theoretical fluid dynamics.

The interpretation of the first of the four equations is: in the U bottle, at time $n+1$, the volume of the water present (U_{n+1}) equals what it was at the nth time (U_n), plus what the pump delivered (P_n), minus what the two tubes drained out ($f_1(U_n)$ and $f_2(U_n)$). Similar interpretations hold for the other three equations.

If we add up both sides of the four equations, we find that many terms cancel, leaving

$$U_{n+1} + V_{n+1} + W_{n+1} + R_{n+1} = U_n + V_n + W_n + R_n, \; n = 1,2,3, \ldots.$$

This is the global *conservation principle*. It says that if there is no overflow catastrophe, the total amount of water in all the bottles and the reservoir must remain constant.

Solution of the Continuity Equations

The continuity equations designated by (*) constitute a system of four "difference equations." To solve them one must know the *initial state* of the system in the first time interval, i.e. the volumes of the water in the individual containers at that time. A typical selection might be: R_1 = total volume of water, $U_1 = 0$, $V_1 = 0$, $W_1 = 0$. This describes an initial condition in which all the water is in the reservoir.

The set of four equations (*) are in a particularly easy form known as triangular form. This means that the first equation involves only one unknown, U_n. The second equation involves only two unknowns, U_n and V_n, etc. On the basis of our knowledge of U_1, we use the first equation to compute U_2. On the basis of our knowledge of U_1 and V_1, we use the second to compute V_2. The third equation is then used to compute W_2 and the fourth to compute R_2. At this stage of the game, we repeat the whole process, and use our knowledge of U_2, V_2, W_2, R_2 to compute U_3, V_3, W_3, R_3. In the same way, we can compute the values at the fourth observation time, and so on as far as we wish.

An explicit set of formulas solving (*) is generally available only if the system (*) is linear, but it is a very easy task to solve it numerically on a computer. This is known as "running the model." With the superb computer graphics that are now available, we could easily use a computer to simulate the whole process and get rid of the bottles and tubes in favor of a graphical display. Software packages are available which make this incredibly easy. If we did this, however, an important lesson would be lost: *ultimately we must deal with the physical world in its own terms* and not in our simplified symbolic terms.

The mathematical model set up in this way is open to criticism. For example, we have discretized time, requiring it to "occur" at discrete values. In addition, we have discretized space, replacing a continuous spread of fluid by a kind of average value at four places.

If there are difficulties with our mathematics, there are also difficulties in setting up the experiment. We should like, ideally, to have steady, laminar flow of water in all our tubes, but we rapidly find that bubbling and turbulent flow may occur. This situation is often delicately unstable, with laminar and turbulent regimes following each other in an oscillating fashion. We may find it is possible for several different laminar regimes to be set up in one and the same tube wherein only a fraction of the diameter of the tube is occupied. We also find that if we use a cheap pump, it may not deliver the water at a steady rate. With some patience and precautions, we may finally measure the quantities required for the mathematical model to an accuracy of five percent or so.

To decide whether this mathematical model is good enough to quantitatively predict the behavior of this physical apparatus would take much more intense experimentation and computation.

Equilibrium Values

If the pump is delivering a constant amount of water P in a time interval, it might occur that each of the quantities U_n, V_n, W_n, R_n approaches a limiting equilibrium value as n goes to infinity. We use the symbol ∞ to denote infinity, and we call these limiting values U_∞, V_∞, W_∞, R_∞. Then, as a moment's thought will make clear, U_{n+1} approaches the same limit as U_n; and similarly for V_{n+1}, W_{n+1} and R_{n+1}. Thus, allowing $n \rightarrow \infty$ in (*) and assuming that all the functions are continuous, one obtains

$$
(**) \qquad
\begin{aligned}
P &= f_1(U_\infty) + f_2(U_\infty) \\
f_1(U_\infty) &= f_3(V_\infty) + f_4(V_\infty) \\
f_5\,(W_\infty) &= f_2(U_\infty) + f_4(V_\infty)
\end{aligned}
$$

This is a system of three equations in the three quantities U_∞, V_∞, W_∞. The last limiting equation derived from (*) is dependent (redundant) and has been eliminated. The interpretation of these equations is that at equilibrium the inflow to each bottle is just balanced by the outflow.

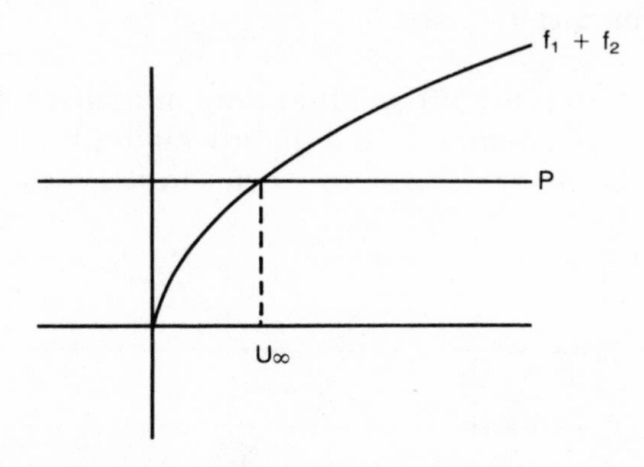

As mentioned before, we may assume $f_1, f_2, \ldots, f_5$ are increasing functions of their variables. To solve the system of three equations (**) first solve the first equation for U_∞. There is a unique solution: see the figure. Using this solution, solve the second equation for V_∞. Using these two solutions solve the third equation for W_∞. The theoretical equilibrium values obtained in this way may be within the capacity of the vessels, in which case equilibrium is physically possible. If they are not, equilibrium is not possible.

Control

To achieve a specified water level or to avert an impending disaster, one turns the pump and the relevant petcocks up or down. Usually, but not always, disaster can be avoided. This is an example of manual control.

Automatic control could be installed in the equilibrium machine, although the equipment is not so easy for the amateur experimenter to come by. In automatic control, a sensing device placed in all the containers senses the level of the liquid. If it is too high (or too low), the flow at the pump or at the petcocks is automatically adjusted. In this case, the functions $P, f_1, f_2, \ldots$ would then depend on the U_n, V_n, W_n, R_n, and the mathematical interlocking expressed by (*) would be complicated indeed. A major problem, both practical and theoretical, would be how can appropriate control be achieved and maintained?

Stability of the Steady State

When a few coins are thrown into the Fountain of Trevi, we expect that the water will equilibrate at a slightly higher level. If the pumping rate were lowered slightly due to a drop in the main line voltage, we

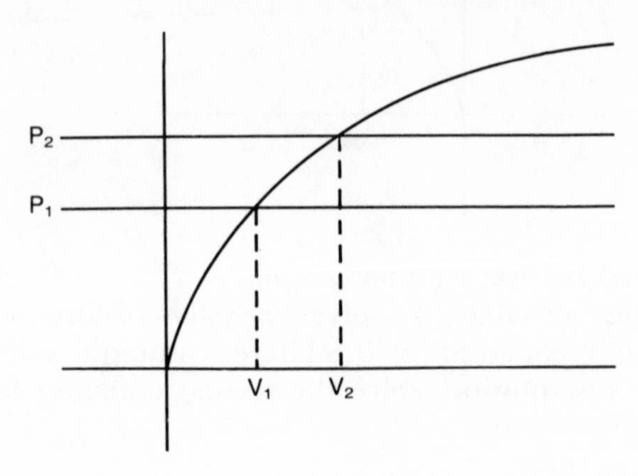

would not anticipate a great catastrophe. In other words, a small change in the physical data produces only a small change in the behavior of the machine. This is called the stability property. One can see this from the graph: if P is moved up slightly from P_1 to P_2, this merely has the effect of moving up slightly the equilibrium position from V_1 to V_2.

Wider Implications

The theory of feedback systems, control, and stability poses some of the most difficult questions in theoretical mathematics and is currently a lively field of research. At the practical level, the generic problem is this: we have a complicated system (the economy, the ecology, the human body, a rocket, etc.); what shall we do in order to bring about such and such a response from the system? This is obviously of vital importance.

The implications of equilibrium, stability, and control on the political and historical scene are obvious. History can be written as the story of the degree to which civilization needs or wants equilibrium, stability, and control and the price it has been willing to pay for them. In cosmology, too, one is interested in the beginnings of the universe (if any) and in its ultimate fate. Periodic revitalization and regeneration? Or the self–obliteration by the equilibrium of total disorganization (thermodynamic heat death)?

When one plays with a real set of bottles, tubes, and pumps, one finds that it is not so easy to achieve the steady state. One turns down petcocks at one bottle and reaps disaster at another bottle. Until one gains a vast amount of experience, a good deal of what happens seems counter–intuitive. One gains, perhaps, a bit more respect and sympathy for the people in government who attempt to improve economic conditions by changing a few financial percentage points.

Mathematical models such as the one just discussed are in widespread use. Some involve a hundred or more equations. They are easy to make, easy to run on a computer, but difficult to construct and adjust to reflect the world accurately. They are full of methodological "maybe's." They are used as Oracles, and public policy may be based on their oracular reputation. They are the best that so–called rational thought has yet provided.

Society perceives a need for certain changes. How are those changes to be achieved without driving the whole system into dis–equilibrium or chaos? By revolution? The world is now so interlocked that there is no guarantee that a revolution undertaken for the most benign reason will not induce undesirable side effects which the authors of that revolution cannot foresee. By actions based on dogma? By rational planning? What is "rational"?

One does what one can *and too often we have to mop the floor.*

Further Readings. See Bibliography

R. Aris and M. Penn; V. Arnold, G. Dahlquist and A. Bjork; F. Hoppenstadt; W. Mayer

Computer Graphics and the Possibility of High Art

THE ORIGINS OF MATHEMATICS are to be found in three activities: counting, measuring, and visual art. Of the three, the first two have received much more attention. We have ancient texts dealing with counting and measuring which constitute a palpable record of primitive mathematics. Although we have innumerable instances of the arts of weaving, architecture, pottery, jewelry-making, and many others, we are only able to infer here an implicit relationship to mathematics. Over the centuries the importance of explicit mathematics to art has waxed and waned and waxed. One suspects that the artist is often an unconscious mathematician, discovering, rediscovering, and exploring ideas of spatial arrangement, symmetry, periodicities, combinatorics, transformations, discovering in an intuitive sense visual theorems of geometry, but periods of conscious mathematization are few.

Renaissance art often had a mathematical quality; the artists were interested in solid geometry, in problems of perspective and of foreshortening. "Let no one who is not a mathematician read my works," wrote Leonardo da Vinci in his *Trattato della Pittura*. Some of Dürer's engravings show the artist working with ruled coordinate frames, so that the hand–eye coordination is reinforced automatically and mathematically.

In recent times, the Bauhaus school has been self–consciously mathematical, though we sometimes wish its members had graduated from the rectangle and the circle. The surfaces of Henry Moore often look more like models sitting in the glass cases of mathematical museums than like coverings for the flesh of humans. The early work of de Chirico is unabashedly mathematical in spirit, as is the later work of Mondrian. The work of Lippold and other artists who work in thread and wire exemplify the principle of tangent lines enveloping space curves.

The artist with the strongest mathematical appeal and general popularity in recent years is undoubtedly M.C. Escher (1898–1972). Escher

Computer rendition of Uccello's (1397–1475) Chalice. (*Courtesy of Computer Aided Geomitric Design. Elsevier Science Publishers B.V. (North-Holland).*)

was an intuitive geometer and, in his later years, worked with professional mathematicians, produced a body of work which won the acclaim of many mathematicians as well as a popular audience. His work has had the accolade of having been transferred to yard goods. Enshrined now in Hofstadter's book *Gödel, Escher and Bach*, he depicts graphically the paradoxes of self–reference and the metaphysical mysteries and amusements that derive from these paradoxes.

Graphical art is, at bottom, a quantity of paint smeared on a canvas or some ink deposited on paper. Through the sense of vision and its interaction with the imagination, these spots and dots are often associated with other objects of the physical or mental world. In the current jargon of computer science, our mind "pops them up" to a metalevel of altered meaning.

Art always draws on the tension between metalevels and between being and non–being. Escher's paradoxes are part of the large kit of illusions and imaginings that are available to the artist. The endless staircase in his picture "Ascending and Descending" elicits spatial confusion

An early mathematization of the Art Process. (*Courtesy of the Lownes Collection, John Hay Library, Brown University.*)

and laughter, and as a consequence it suffers from the general aesthetic judgment which tends to place pun, wit, comedy, irony at a lower level than pathos, tragedy, awe, or intimations of the transcendent.

The aesthetics of Escher's "Sky and Water" (both I and II) is rather more subtle. Here one has a regular division of the plane, a group of subtle transformations of the figure into itself, a clever use of a visual pun wherein the space between the basic figures is itself organized into a subsidiary figure of increasing importance.

This kind of intellectuality appeals greatly to the mathematician who sees in the figure an application of group theory and analyzes the figure into glides, reflections, reversals, rotations, etc. To the mathematically unsophisticated, the picture makes the visual appeal of all patterned work that exhibits symmetry groups. It is further enhanced by its stunning execution and an underlying emotional charge derived from an apparent release from mathematical rigidity into freedom. Though this picture is a masterpiece of imagination and execution, it fails ultimately as high art, precisely because the mathematical element dominates. Its spatial harmonies are, in the end, like harmonies of crystals: part of inanimate nature, scintillating but emotionally unsatisfying.

Inevitably, the draftsman and the artist met the computer scientist. The equipment for computer graphics is currently of several types. There are plotters wherein a pen or a stylus is moved across a thin

Sky and Water I, M.C. Escher. (*Copyright M.C. Escher Heirs c/o Gordon Art, Baarn, Holland.*)

sheet of material. There are line printers and squirt printers wherein tiny symbols or blobs of ink are deposited line by line on paper. There are scopes, both in black and white and in color, wherein the scope face is subdivided into tiny areas called "pixels" (1024 × 1024 pixels on the scope I'm familiar with) and the electron gun illuminates prescribed pixels in a prescribed order. The scope face may then be photographed for hard copy or a video tape prepared for distribution and playback. For genuine three dimensional graphics there are lathes and cutting devices which work on a variety of materials. All of these image–creating machines are driven by a computer, properly programmed.

The graphical requirements of a scientist are very often nothing more than a picture of a mathematical function of one variable: $y = f(x)$, displayed with axes labelled with an explanatory legend. Sometimes a function of two variables $z = f(x,y)$ is wanted, and this can be displayed as a sequence of one dimensional curves, or as a projection of a two dimensional surface onto a plane. Occasionally the graphical needs of science go beyond this, and one may have to put together by combination of simple elements very complicated pictures such as molecular arrangements in space.

Computer Aided Geometrical Design (CAGD) is a new branch of mathematics–computer science that creates and manipulates surface shapes appropriate to industrial production.

A very recent development in graphics, of growing importance in medicine, is the reconstruction and display of the interior of portions of the body on the basis of information received from probes. This field is known as tomography.

The mathematicians who have created pictures of fractal dimension, who have revealed unsuspected visual worlds lurking behind simple formulas, the image processors who have plied their Fourier techniques, the vision experts who inform us of what we see at the edge of an object, have all contributed to a computer graphics of stunning variety and visual quality.

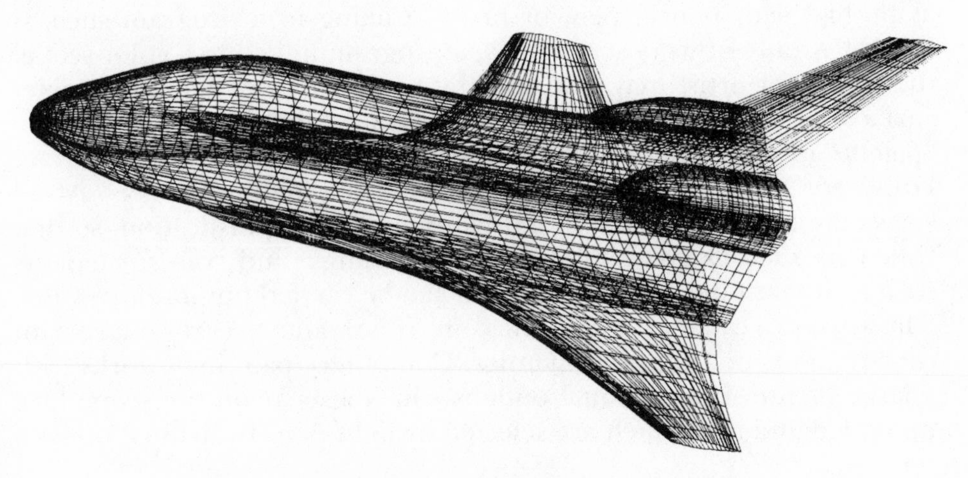

Computer aided geometrical design. (*Courtesy of NASA.*)

The line printer and the computer scope were hardly in place when computer graphics took off from its role as a useful adjunct to science, technology and business, and computer art was born.

Donald Ducks, Abe Lincolns, and pinup girls were spewed out by the teletype machines of the early fifties. By the mid 1960s the public was presented with a flow of graphical material created by means of the computer and was invited to judge it, react to it, and enjoy it, like conventional graphics. Courses were offered in computer art. Gallery shows were held but were condescendingly reviewed, as though this were a medium without serious intent or noble aspiration. Computer artists, people whose talents and intentions combined the mathematical, the programmatic, and the artistic, were held in contempt by each of the pure establishments, and starved as they sat in garrets with their second–hand plotters.

In the early stages of computer art, the formal (as opposed to intuitive) mathematical requirements for its practitioners were substantial. At that time the basic visual components were specified mathematically: a certain curve or shape, a certain method of transforming, juxtaposing, and iterating these components, and programming was done in low to intermediate level computer languages. By the mid 1970s the push was on to move the craft to a higher level of computer language and to free the artistic component from the mathematical and programmatic components. With the successful introduction of the light pen, the "sketchpad" tablet, and the joystick, and with a high level language in place, the artist could function almost—but not quite—as he had with pencil, pen, or brush. Calling in a program such as PAINT or any of its descendants*, a program linked to a color scope, the computer artist may draw on the scope surface, or on the tablet surface, in a conventional manner. He may "dip" his "brush" into a "palette" of colors, indicating his desired hue, intensity and saturation. Hundreds of colors are available and are on "menu display." At all stages the artist has a software backup of great sophistication, so that functions such as color fill, texturing, scalings, and transformations such as replication and juxtaposition can be carried out automatically. The artist can call in a "flip of the coin" if he wants to carry out certain constructions in a random manner. Computer "painting" works with a large menu of operational options which appear on the scope face upon demand and which are selected by light pen. Individual options

* Such programs are now widely available on personal computers.

R. Vitale rendered by parametric interpolation. (*Artists: Ruth and Keith Long, 1973.*)

may have sub–options to two or three sub–levels. It is fascinating to watch a skilled computer artist working with a PAINT program and to compare the hand–eye coordination with that of painting with acrylics or oils.

The 1970s also saw an increasing amount of computer animation. It is very easy to program minute changes of a base figure, and when changes are displayed at the proper frequency the eye perceives continuous motion. The figures on the scope become alive, and by photographing successive displays one produces animated films. In the case of humanlike figures, it takes some mathematical subtlety to have the computer interpolate automatically between an initial figure and a final figure, but this has been worked out fairly successfully now. The Canadian Film Board produced some computer–animated films of unusual quality.

By the late 1970s computer animation had produced so many examples of quality that a weekly TV show devoted to it became possible. Now computer graphics and animation, combined with conventional animation techniques, photography, and special effect photography, have become commonplace in commercials, videogames, and science

Computer Bugs: An in-house, high level scene description language was used exclusively to model this image on a VAX 11/780 running 4.2 BSD Unix. Original in gorgeous color. (*Courtesy of the Department of Computer Science, Brown University. Artists: Ben Trumbore and Kelan Putbrese.*)

fantasy films, and one hears that a great deal of development goes on behind the closed doors of commercial studios.

The appearance of a new technological medium opens a variety of questions. What reasons can be given for working in the new medium? What new kinds of things are created? What is their ideational, emotional, or social impact? To what extent is computer art created automatically or serendipitously by the machine, and to what extent is it all under the precise control of the artist? What is the relationship between the craft process and the final product? What are the prospects for the future? Are masterpieces possible? Finally, returning to our opening theme, what is the relationship between mathematics and fully matured computer graphics?

To both those interested in creating art or criticizing it and those who are interested in semiotics, the idea of creating visual effects through a formalized symbolic process executed via computer is an attractive one. Students of pattern and sign may inquire about the role of a high level grammar of visual nouns and verbs as opposed to atomic spots, dots and impastos. One might get a certain kind of chair from software, very much as a stage designer gets a stuffed Victorian loveseat from a supplier of stage properties. (He probably does not construct it in his shop.) To animate a horse, one might have available in software the standard verbs "to trot," "to canter," "to gallop." At a very high level, of course, one has always used generic terms: landscape, still life, portrait. In the Renaissance a church might have ordered a picture by a higher level designation: an Annunciation, a Descent from the Cross, a Pieta, an Assumption, in full confidence that this description would suffice to provide pretty much what the client wanted. Computers give us the possibility of doing this at every level down to the pixels themselves.

Trompe–l'oeil has existed in art for centuries. High tech now gives us the ability to simulate or to forge reality at an incredible level of visual fidelity. Art is always illusion, and when illusion is accepted as

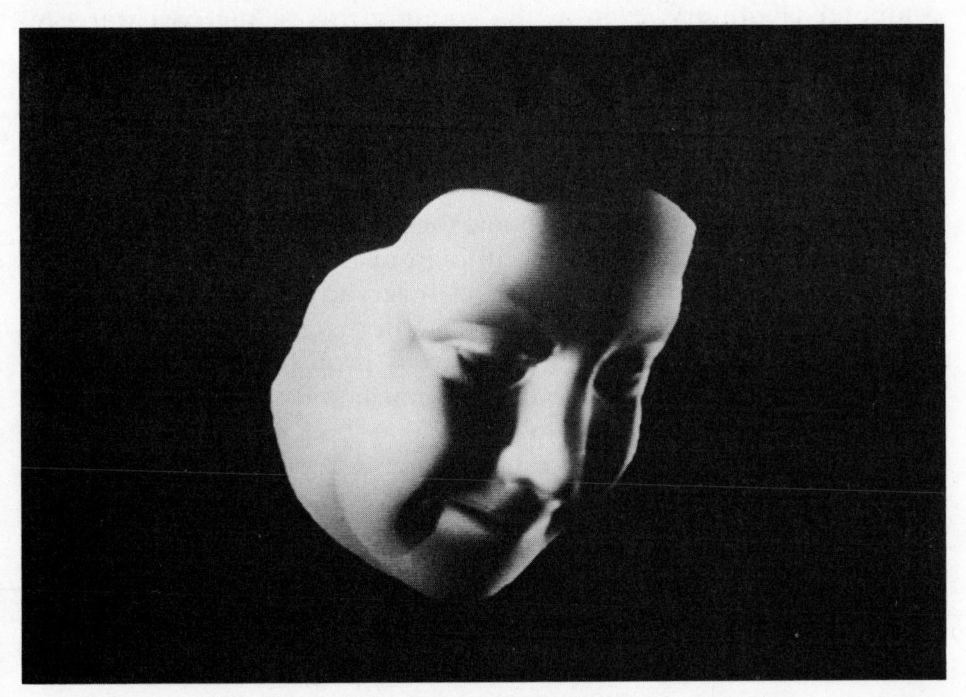

A face constructed with B-Spline surfaces of the fourth order. (*Computer artists: Herve Huitric and Monique Nahas at Regie Renault, France, 1984.*)

reality, it forces us to reconsider the fundamental philosophical question: how do we know what is?

At the present stage of the technology it is hard to separate the product from the technique. The whole software component is still in flux. The artist himself is establishing and stabilizing programs. As programming is open–ended, there is conflict for the artist; should he devote all his time to creating art in the mode already available, or should he develop additional structures of software against present and future needs? Computer art now presents the artist with the challenge of operating with a new medium and of simultaneously expanding greatly this very medium.

The nature of computer art is often misconstrued by the public. You don't just push the button and have art pop out mindlessly. The artist works with a computer which is itself the end result of decades of rational planning. He works in one or several computer languages which have evolved as natural languages have evolved. The artist must have some conception (not necessarily visual) of what to create and how he proposes to do it. The artist serves an apprenticeship in a computer laboratory, even as the young artists of the past were apprenticed to their masters. The artist does not have full control of his product, for, in the passage from the symbolic and programmatic to the visual, he cannot anticipate all aspects of what the machine will create. (Indeed, in the area of scientific computation, this is precisely the reason for using the computer: the human mind cannot anticipate what the formulas imply.) Creation therefore proceeds by successive approximations, guided by the intuition of the artist. The correction process is probably easier than with oil or acrylic, and many preliminary steps may be passed through before a work is accepted as final. The serendipitous element in computer art is probably stronger than in conventional art. Mistakes are not just smears of paint but may result from boo–boos in programming. Such boo–boos can result in graphical surprises of unusual interest.

The very high level languages now used in painting simulation have been deplored by some computer artists. Why should we employ a medium costing hundreds of thousands of dollars to arrive at conventional effects when paint and canvas are so much cheaper? Although there are advantages having to do with color, with the automation of rote processes, with the manipulation of new graphical vocabularies, with correction and modification, with storage, with reproduction, it must be admitted that the complaint has a certain validity. I find that instances of static computer art put forward as art for its own sake have limited appeal. Sometimes an initial reaction of elation, shock, mystery, whatever, comes from the unusual texture or color, from the juxtaposition of elements, or from the creation of superreal objects. Often the underlying iconography hints at a strange and wonderful world of the future that will be brought about by science and technology, a message which after two hundred years is rather trite.

On the other hand, if computer art has a future as an art form in its own right, it is to be found in the dynamic, the animated, the interactive. It should look not towards Rembrandt, but towards Verdi's "Aïda". Not just the classical "Aïda" but an "Aïda" with the audience singing along and scrambling onto the backs of the elephants on stage. Chaos? No. Total theatre.

One has intimations of what is possible in the dynamic, abstract, mind–blowing computer–generated films of pure kaleidoscopic quality. One has further intimations in computer–assisted animated films. One is still closer to future masterpieces, I think, in the video games that are in every bus station and on most home televisions. These popular amusements, replacing older pinball games, combine the dynamic, the visual, the kinaesthetic, the interactive, the rational, the artistic. There is emotion and even a sense of theatre in them, even though it is a theatre of violence.

We live in a world of motion. We are obsessed by it. We are in and out of cars constantly. We jet around the world; we watch TV and movies. We have allowed our cars to destroy conventional architecture and replace it with an architecture for and of mechanized nomadism. In a sense, all static art leaves us unsatisfied; it is incomplete. We may put it on the wall for a splash of color, trudge by it virtuously in a gallery, philosophize about it, buy it and sell it and speculate with it, but our love for it is not deep. Motion, action, transformation, change, a sense of becoming rather than of being, is what quickens our blood.

The medium of computer graphics is young. It is hardly out of the French caves. High art? Masterpiece? Why not? The computer Verdis of the future will captivate us.

Further Readings. See Bibliography

C. Adcock; K. Clark (1979, 1983); J. Foley and A. Van Dam; F. Malina; D. McCloskey; L. Myer; A. Noll; D. Pedoe; SIGGRAPH; G. Youngblood

II

THE SOCIAL TYRANNY OF NUMBERS

Mathematics 163
MWF 10 Sec. 3B 301 Manse
Prof. R.B. Smith TA: F. Jones
Final Exam

Student ID	Grade
072-36-7345	78
140-47-7262	75
149-87-4850	88
241-01-5033	62
362-22-8625	91
384-98-9098	75
509-15-5143	94
522-17-1276	88
791-35-0107	79
798-45-6063	55
807-89-0229	72
936-01-3145	85
987-03-2678	82

AV = 78.769231
SIGMA = 10.821303
MEDIAN = 79.000000

Mathematics and Rhetoric

Introduction

If rhetoric is the art of persuasion, then mathematics may seem to be its antithesis. This is believed not because mathematics does not persuade, but rather because it seemingly needs no art to perform its persuasion. The matter does it all; the manner need only let the matter speak for itself.

In Euclid we find only bare statements of the "common notions" (the "axioms" or "postulates") followed by an unmerciful chain of theorem, proof, theorem, proof. Indeed, in the high–school geometry in which Euclid was force–fed to uncounted millions of school children, "proof" was reduced to a formal scheme, in which two adjacent columns, "statements" on the left and "reasons" on the right, led inexorably from the "given" to the "to prove," from hypotheses to conclusion.

From the definitions, the axioms, and the figure, the theorem is inescapable. Anyone who understands its terms will agree to its truth; if a student should fail to agree, he would thereby declare himself incompetent before his class and his teacher.

"Mathematical certainty" is a byword for a level of certainty to which other subjects can only aspire. As a consequence, the level of advancement of a science has come to be judged by the extent to which it is mathematical. First come astronomy, mechanics, and the rest of theoretical physics. Of the biological sciences, genetics is top dog, because it has theorems and calculations. Among the so–called social sciences, it is economics that is most mathematical, that offers its practitioners the best job market as well as the possibility of a Nobel prize.

Mathematization is upheld as the only way for a field of study to attain the rank of a science. Mathematization means formalization, casting the field of study into the axiomatic mode and thereby, it is supposed, purging it of the taint of rhetoric, of the lawyerly tricks used by those who are unable to let facts and logic speak for themselves. For those who want to assert the claim of rhetoric as a necessary and valid aspect of any human endeavor, mathematics appears as the dragon which must be slain.

Now, my purpose here is to undermine these claims for mathematization. I say undermine, not refute or destroy, for we are well aware that the claims for mathematization are not made without reason, but their validity is limited. A skeptical look should be cast upon mathematical theories as much as upon theories stated in "ordinary language."

My goal is to show that mathematics is not really the antithesis of rhetoric, but, rather, that rhetoric may sometimes be mathematical, and that mathematics may sometimes be rhetorical. My first task will be to show that mathematical language, mathematical trappings, are being used as a rhetorical device in various fields of application and especially in the so–called behavioral sciences. My second task will be to show that within the practise of mathematics itself, among the professional mathematicians, continual and essential use is made of rhetorical modes of argument and persuasion in addition to purely formal or logical procedures. Accordingly, this introduction is followed by Parts 1 and 2 and finally by a brief closure.

Part 1: Mathematics as Rhetoric

There are three branches of mathematics: pure mathematics, applied mathematics, and rhetorical mathematics.

Pure mathematics is number theory, or geometry, or algebra, or analysis. It's what mathematicians do to please themselves, or each other. When they are pleased with the way something comes out, they are likely to say that it's elegant, or deep.

Applied mathematics is what mathematicians do to accomplish the tasks set by the rest of society. It is numerical weather prediction, or statistical quality control of electric light bulb manufacture, or plotting of the trajectory of a rocket to Saturn. More and more often these days, the tasks are set and paid for by the military and involve the preparation of the premature end of life on this planet.

Finally, there is rhetorical mathematics. What is that? It's what is neither pure nor applied, obviously. Not pure, because nothing of mathematical interest is done, no new mathematical ideas are brought forward, no mathematical difficulties are overcome; and not applied, because no real–world consequences are drawn. No practical consequences issue from rhetorical mathematics—except publications, reports, and grant proposals. The word rhetoric means many things. One of its invidious meanings is empty verbiage or pretentious obfuscation. Mathematics can be rhetoric in this sense of the term. We call it rhetorical mathematics.

For example, you might develop a "mathematical model" for international conflict. The model might be just a list of axioms: an axiomatic model. Or it might be a collection of strategies with an associated payoff matrix: a game–theoretic model. Or again, maybe a collection of "state variables" to specify the international military–political situation, together with a set of equations relating the values of the state variables today to their values tomorrow. Program this into your computer, and you've got a simulation model.

It doesn't really matter which way you do it. You can calculate, publish, readjust your model (or throw it out and start again from scratch), calculate again, and publish again.

Why is this activity not applied mathematics? The standard picture of applied mathematics, which can be found in the first few pages of many textbooks, breaks down the work of the applied mathematician into three phases, which can be represented schematically in the following arrow diagram:

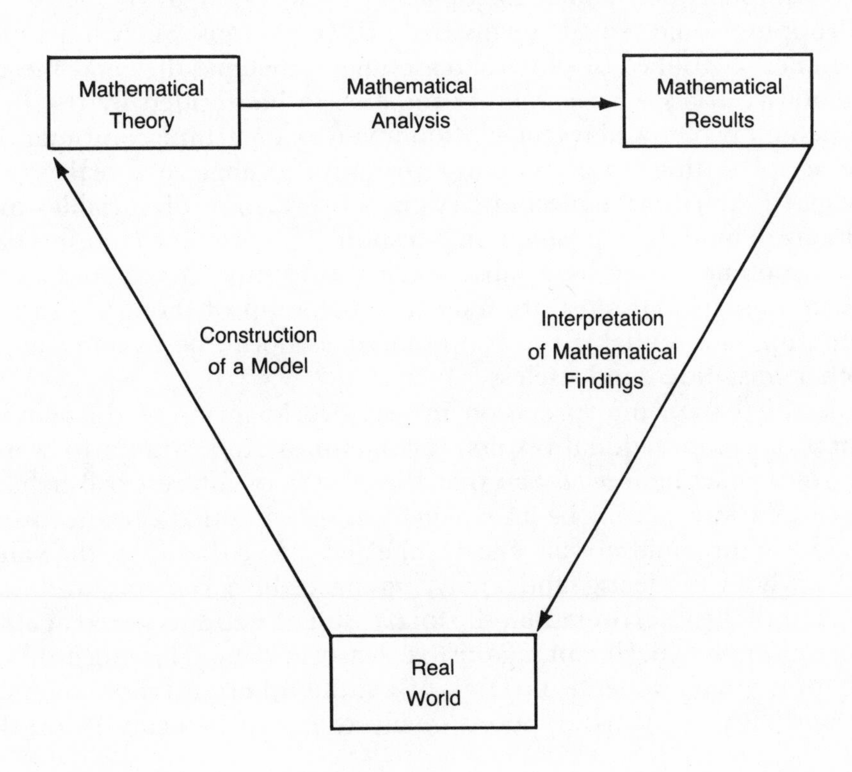

The upper level is theory; the lower level is physical reality. The mathematical study of real–world problems (as distinct from problems in pure mathematics) begins by construction of a mathematical model. This means the representation of quantities of physical interest by mathematical variables (most often numerical, but sometimes non–numerical, for instance, geometrical or logical) and the derivation from physical experience of relations among these variables (most often, algebraic and differential equations and inequalities).

The second step, mathematical analysis, may in some cases be *solving the equation*, obtaining an explicit formula for some variable, such as temperature, population size, or position of a planet. In other cases, an explicit solution may not be attainable, but some approximate or qualitative conclusions can be obtained by mathematical reasoning; for example, the planet will remain within a certain distance of the sun; or, the population will at first increase rapidly and later level off and approach a certain limiting value; or, the temperature depends smoothly and monotonically on the diffusivity of the medium.

In addition to strict mathematical reasoning, step two may involve *ad hoc* simplifications, such as replacing some variables by constants, or dropping some "small" terms from the equations. Such steps may sometimes be justified by physical reasoning; sometimes they may merely be tentative trials, whose validity remains to be decided by the final result. Step two nowadays most often involves a machine computation. The act of setting up a computer program to analyze a real–world problem requires as a preliminary the introduction of variables and relations to model the problem in question. The machine computation may sometimes serve as a labor–saving substitute for thought and human analysis, but most often a certain amount of thought prior to computing is essential if the computation is not to be in one way or another misguided and useless.

The third step, interpretation in real–world terms of the mathematical or computational results, is sometimes quite straightforward. The result may be a *prediction* that the system of interest will behave in a certain way. It may be an *explanation*, showing that certain causes could have (or could not have) certain effects. In either case, the value of the whole modeling and analyzing procedure remains undetermined until the interpretation, the final result of step 3, is tested against observation or experiment, against real–world data. The merit or validity of a model depends on, first of all, the inherent reasonableness or plausibility of the assumptions involved in step 1; secondly, on the tractability of the model, the possibility of carrying out in step 2 some mathematical operations leading to conclusions of some novelty and interest; and finally, in step 3, on the goodness–of–fit of the results, on the degree to which the theoretical results conform to the real–world data.

This "Schaum's outline" of scientific methodology is intended to give criteria by which one may evaluate the claims for applications of mathematics in one or another field of study. The three–step paradigm is conventional and perhaps simplistic. Any particular piece of research may be limited to only one step of the three, or all three steps may be iterated several times as a model is gradually refined and corrected. It may sometimes be impossible or inconvenient to make a clear–cut demarcation between one step and the next.

Granted all this, there remain certain criteria by which the mathematician judges whether an "application" of mathematics is genuine or bogus.

Does the depth of the real–world problem justify the complexity of the mathematical model?

Are any genuine mathematical reasonings or non–trivial calculations carried out which require the resources of the mathematical model being proposed?

Are the coefficients or parameters in the equations capable of being determined in a meaningful and reasonably accurate way?

Are the conclusions capable of being tested against real–world data? Do any non–obvious practical conclusions follow from the analysis?

The introduction of mathematical methods in biology, economics, psychology, and other branches of the so–called behavioral sciences has always been accompanied by controversy. The opponents to mathematization may have had good grounds for their resistance, but their arguments could be discounted by raising the suspicion that they didn't understand the mathematical methods they were challenging.

For this reason, it is important to state publicly that among professional mathematicians the skepticism about behavioral–science mathematics and even about mathematical biology is much stronger than it is among non–mathematical behavioral scientists and biologists.

This skepticism is rarely stated in print. Unlike philosophers and literary critics, mathematicians dislike controversy. They are not used to it and will usually keep their mouths shut to avoid it. (A famous instance was Gauss' suppression of his own discovery of non–Euclidean geometry, for fear of a clamor from the "Boeotians.")

An additional reason why mathematicians seldom state in print their skepticism about behavioral–science mathematics is that we know that *some* of it must be worthwhile, so one cannot condemn *all* of it. As a practical matter, to separate the wheat from the chaff would be a dreary undertaking. As a consequence, we say nothing, but, behind the back of the speaker on mathematical psychodynamics, we raise our eyebrows at each other and shrug.

Perhaps the knowledge that mathematicians share their opinion will strengthen the resolve of those who wish to oppose rhetorical mathematics. If you need advice or encouragement from a professional mathematician, you need go no farther than the mathematics department on your own campus. Look for the best mathematician you can find. It doesn't matter if he's pure or applied, what matters is that he has high mathematical standards.

Perhaps our negative definition of rhetorical mathematics should be restated in positive terms. Rhetorical mathematics is a form of academic gamesmanship. It relies above all on the high prestige accorded mathematics by twentieth–century North America. Rhetorical mathematics presents itself as applied mathematics, but it is easy to tell them apart. Applied mathematics sooner or later leads to an experiment or a measurement. Either initially or ultimately, work in applied mathematics leads back to the phenomenon being modeled. Rhetorical mathematics is often incapable in principal of being tested against reality. For instance, the model may contain numerical parameters that are obviously incapable of measurement. (E.g., a model of international conflict, with coefficients equal to the "aggressiveness" of the major powers.)

An amusing example is brought to light in Neal Koblitz's essay, "Mathematics as Propaganda." He quotes from *Political Order in Changing Societies*, the definitive work on problems of developing countries by the very influential Samuel Huntington. On page 55 of this book are found three equations relating certain social and political concepts:

$$\frac{\textit{social mobilization}}{\textit{economic development}} = \textit{social frustration}\ (\frac{a}{b} = c)$$

$$\frac{\textit{social frustration}}{\textit{mobility opportunities}} = \textit{political participation}\ (\frac{c}{d} = e)$$

$$\frac{\textit{political participation}}{\textit{political institutionalization}} = \textit{political instability}\ (\frac{e}{f} = g)$$

As Koblitz remarks, "Huntington never bothers to inform the reader in what sense these are equations. It is doubtful that any of the terms (*a*)–(*g*) can be measured and assigned a single numerical value. What are the units of measurement? Will Huntington allow us to operate with these equations using the well–known techniques of ninth grade algebra? If so, we could infer, for instance, that

$$a = bc = bde = bdfg \,,$$

i.e., that "social mobilization is equal to economic development times mobility opportunities times political institutionalization times political instability!"

A more notorious example is the book *Time on the Cross* by Robert W. Fogel and Stanley Engerman. This book created a sensation in 1974. Using statistical arguments with computer–processed data, it purported to show that the slave system in the pre–Civil War South was more humane and more efficient than the free labor system in the North at that time. A horde of critics soon discredited the claims of that book. Some particularly penetrating comments were made by Thomas L. Haskell in the New York Review of Books, and are reproduced in Koblitz' article. Haskell considers *Time on the Cross* as an exemplar of "cliometrics," the use of quantitative methods in the study of history.

"On the surface, cliometrics is an austere and rigorous discipline that minimizes the significance of any statement that cannot be reduced to a clear empirical test ('operationalized'). But beneath the surface one often finds startling flights of conjecture, so daring that even the most woolly–minded humanist might gasp with envy.

"The soft, licentious side of cliometrics derives, paradoxically, from its reliance on mathematical equations. Before the cliometrician can use his equation to explain the past, he must assign an empirical value to each of its terms, even if the relevant empirical data have not been preserved or were never recorded. When an incomplete historical record fails—as it often does—to supply the figures that the cliometrician's equations require him to have, it is considered fair play to resort to *estimation*, just so long as he specified the assumptions underlying his estimates. And although cliometrics requires that these and all other assumptions be made explicit, it sets no limit at all on the *number* of assumptions one may make, or how high contingent assumptions may be piled on top of each other—just so they are explicit."

Part 2: Rhetoric in Mathematics

We turn now from rhetorical mathematics to mathematical rhetoric. We want to look at mathematical utterances or writings (the talk or writing of mathematicians in the pursuit of their work as mathematicians) and see what rhetorical aspects we can identify.

On the basis of the customary definition of rhetoric as natural discourse which serves to convince, rhetoric in mathematics would simply be common language put to the purpose of convincing us that something or other about mathematics is the case. What might we want to argue rhetorically? Certainly we would want to argue the utility of mathematics in its many applications. The philosophy of mathematics is also built up by rhetorical argumentation. But the *truth* of mathematics—moving down one level from a discussion of the truth to the truth itself—is considered to be established by means which are the antithesis of rhetoric. The claim made in the classroom, in the textbook, and in a good deal of philosophical writing, is that mathematical truth is established by a unique mode of argumentation, which consists of passing from hypothesis to conclusion by means of a sequence of small logical steps, each of which is in principle mechanizable. T.O. Sloane has written ("Rhetoric," Encyclopaedia Britannica), "All utterance, except perhaps the mathematical formula, is aimed at influencing a particular audience at a particular time and place." Mathematical utterances, it would seem, stand apart. But the small measure of doubt that Professor Sloane has allowed himself can be greatly enlarged. Mathematical proof has its rhetorical moments and its rhetorical elements.

Suppose you were to eavesdrop on a college mathematics class which is sufficiently advanced that the instructor sets considerable store by mathematical proof. Imagine that you have broken into the lecture in the middle of such a proof. Ideally, as we have said before, you should be hearing the presentation of those small logical transformations which are to lead inexorably from hypothesis to conclusion. To some extent you will hear the recital of such a litany but other phrases will inevitably intervene: "It is easy to show that . . . ," "By an obvious generalization . . . ," "a long, but elementary, computation, which I leave to the student, will verify that . . . ."

These phrases are not proof: they are rhetoric in the service of proof. A hilarious compendium of rhetorical devices, used as proof substitutes, has recently been circulating among graduate students in mathematics and computer science. We quote a few lines from this work which was written by Dana Angluin of the Yale Computer Science Department.

How to Prove It

Proof by example: The author gives only the case $n = 2$ and suggests that it contains most of the ideas of the general proof.

Proof by intimidation: "trivial".

Proof by eminent authority: "I saw Karp in the elevator and he said . . .".

Proof by cumbersome notation: Best done with access to at least four alphabets and special symbols.

Proof by semantic shift: Some standard but inconvenient definitions are changed for the statement of the result.

Proof by exhaustion: An issue or two of a journal devoted to your proof is useful.

Proof by metaproof: A method is given to construct the desired proof. The correctness of the method is proved by any of the techniques of this list.

.

.

.

[And so on for a total of twenty–four different categories.]

It is believed by many that all these rhetorical handwavings, desk-poundings, appeals to intuition, to pictures, to the lack of counter–evidence, to meta–arguments, to the results of papers which have not yet appeared, reflect only the laziness of the lecturer or author. Somewhere behind each theorem that appears in the mathematical literature, there should (must) stand a sequence of logical transformations moving from hypothesis to conclusion, absolutely comprehensible, certified as such by the authorities in the field, verifiable as such by even the tyro who knows only the logical moves, and accepted by the whole mathematical community. This impression is absolutely false, yet it is commonly held by people outside the mathematics profession. Mathematics students sometimes carry this picture in their minds until they are themselves involved in research; at this point they experience a sudden and unexpected shock when they realize that the real world of mathematics is far from the ideal world.

In the real world of mathematics, a mathematical paper does two things. It testifies that the author has convinced himself and his friends that certain "results" are true, and it presents a part of the evidence on which this conviction is based.

It presents part, not all, because certain "routine" calculations are deemed unworthy of print. The reader is expected to reproduce them for himself. More important, certain "heuristic" reasonings, including perhaps the motivation which led in the first place to undertaking the investigation, are deemed "inessential" or "irrelevant" for purposes of publication. Knowing this unstated background motivation is what it takes to be a qualified reader of the article.

How does one acquire this background? Almost always, it is by word of mouth from some other member of the intended audience, some other person already initiated in the particular area of research in question.

What does it mean for a mathematician to have convinced himself that certain results are true? In other words, what constitutes a mathematical proof as recognized by a practicing mathematician? Disturbing and shocking as it may be, the truth is that *no explicit answer can be given.* One can only point at what is actually done in each branch of mathematics. All proofs are incomplete, from the viewpoint of formal logic.

How do we decide which of these incomplete proofs are wrong, and which are correct, in the sense that they are convincing and acceptable to qualified professionals? This can be answered only by mastering the mathematical theory in question. The answer involves knowing the difference between a serious difficulty and a routine argument. A mathematician who is a certified expert in algebraic number theory might be quite unable to tell a correct from an incorrect proof in nonstandard analysis.

All that one can say is that part of being a qualified expert in, say, algebraic number theory is knowing which are the crucial points in an argument where skepticism should be focussed, which are the "delicate" points as against the routine points in an argument, which are the plausible–seeming arguments that are known to be fallacious.

A mathematics research article (or reference work or treatise) is *never* written out in complete logical detail. If it were, no one would want to or be able to read it. Its logical completeness would not make it more comprehensible, rather it would make it incomprehensible, except perhaps to computing machines. (We return to the computing machine angle a few pages below.)

If it is not completeness in the sense of formal logic, then in actual practice what does guarantee correctness of mathematical proofs?

Well, there is the referee, or referees, whose approval is a necessary condition for publication. Do the referees fill in and check all the logical details of every argument? Not at all. After all, they are busy people, and refereeing is done free, as a service to the profession, on top of all their other duties. It would be difficult to obtain any broad picture of what referees actually do, since this is an activity that is private and semi–anonymous (known only to editors). Certainly there is a tremendous variation in referees. Some read every line and check every calculation; they refuse to referee any paper they cannot check in this way. It is our impression that only a small percentage of the papers published in mathematical journals receive this kind of refereeing.

For one thing, only another mathematician whose interests and training are very close to the author's would be willing and able to do this kind of checking. Such a referee would likely be favorably prejudiced toward the submitted article; he might be a poor judge of its interest and importance for the mathematical community at large. Someone more detached from the author's special interest might be more objective but probably less intensive in his reading.

A well–known American probabilist once described the refereeing process as follows: "you look for the most delicate part of the argument, check that carefully, and if that's correct, you figure the whole thing is probably right."

Undoubtedly, other factors also will influence the referee's judgement. Do the methods and result "fit in," seem reasonable, in the referee's general context of his picture of the field? Is the author known to be established and reliable, or is it an unknown, or, worse still, someone known to be unoriginal or liable to error?

If an article appears in print, it is hard to be sure what that means, in terms of anyone but the author's having thoroughly understood its contents. It might help if one knew at first hand the editorial and refereeing policies of the journal in question. An editor has been quoted to us to this effect: "By choosing the referee in one way or another, I can guarantee that any particular article will be either accepted or rejected."

Once an article is published, it might be thought that it is subject to the scrutiny of the whole mathematical community. Far from it. Most published mathematical articles attract very few readers and are forgotten within a few months, except by their author and perhaps the author's graduate students.

There are, of course, articles which are widely read and influential. "Widely read" must be understood in a relative sense; in most mathematical specialties, the total active practitioners (publishers of research articles) is at most a few hundred or so, possibly much less. The results that appear in an influential article will be read by dozens or scores of people and will be presented in seminars across the country and around the world. There is a reward waiting for the student or mathematician who can find a serious error in such a paper. There is also a reward for finding extensions, generalizations, applications, alternative proofs, connections with other results.

If a mathematical result attracts widespread attention and survives continued scrutiny and analysis, it enters what might be called the tried and tested part of mathematics.

Does it then have guaranteed certainty? Of course not. The geometry of Euclid was studied intensively for 2,000 years, yet it had major logical gaps that were first detected in the 1880s. How could we ever be sure that we are also not blind to some flaw in our reasoning?

Aha, someone may answer, we could be sure if we would only take the trouble, however troublesome it might be, to code our mathematical proofs in some appropriate computer language, insist that proofs be restricted to logical steps whose conditions could be incorporated in a computer program, and thereby make our proofs verifiable by machine.

As a matter of fact, this idea has actually been tried. One of the most arduous efforts in this direction was carried out in the 1970s by the Dutch mathematician N.G. de Bruijn and his associates. They developed a special computer language, AUTOMATH, with an associated AUTOMATH program. Their goal was to automate the process of checking the correctness of mathematical proofs. After years of intensive experimentation, the Automath project has been virtually abandoned. There are several reasons for this.

1. The formalized counterpart of normal proof material is difficult to write down and can be very lengthy.
2. Even if these translations into Automath were available in great abundance, how would one verify that they were correct, that the Automath program is itself correct, that the machine program has been correctly written, that it all has been run correctly?

3. Mathematicians and computer scientists are not really interested in doing this kind of thing.

The Automath approach represents an unrealizable dream. At the turn of the century, one might have said that a proof is that which is verifiable in an absolutely mechanical fashion. Now that a much more thoroughgoing mechanization is possible, there has been a reversal, and one hears it said that computerizability is not the hallmark of a correct proof. Through all this, the accepted practice of the mathematical community has hardly changed, except for the enlargement of the computer component.

In recent years, researchers at the University of Texas, at Argonne National Laboratory, and elsewhere have made major progress in "automated reasoning" or "automated theorem proving." Their goal, however, is not to establish certainty for mathematical knowledge, nor even to increase the reliability of mathematical proofs, although the latter might come as a side benefit. Instead, their goal is to provide mathematicians with a logical "assistant" comparable to the numerical "assistants" now available for solving algebraic or differential equations.

The "automated theorem prover" is not expected ever to be a "black box" into which the mathematician could drop a few axioms and then wait until it extrudes a theorem. Rather, it would be used interactively; that is to say, it would be an exploratory tool. The mathematician would say to it: "try this hypothesis and this search strategy, calculate for five minutes, then come back and show me what you find."

This orientation on the part of researchers into automated theorem proving is in keeping with the usual attitude of professional mathematicians toward formal logic. Namely, it is just another tool with whose help we attempt to discover mathematical truth. While, in part, logic does control the mathematician's intuition, it is at the same time subject to that intuition. The "automated reasoning" researchers share our skepticism toward the ancient ideal that mathematical knowledge can expect to attain absolute certainty.

We can show the difficulties of "formalism" at an elementary level by looking at an attempt to present a complete, rigorous proof of a very simple theorem.

Even for a very tiny piece of mathematics, the task of giving an absolutely air–tight formal proof turns out to be amazingly complicated. Professedly rigorous proofs usually have holes that are covered over by intuition. Consider the example displayed below.

A Simple Example of a Deductive System

The primitive terms are person and collection.

Definitions. A *committee* is a collection of one or more persons. A person in a committee is called a *member* of that committee. Two committees are equal if every member of the first is a member of the second and vice versa. Two committees having no members in common are called *disjoint* committees.

Axioms. 1. Every person is a member of at least one committee.
2. For every pair of persons there is one and only one committee of which both are members.
3. For every committee there is one and only one disjoint committee.

Theorem. Every person is a member of at least two committees.

Proof.

Statement	Reason
1. Let p be a person	hypothesis; naming
2. p is a member of some committee C	Axiom 1; naming
3. Let D be the committee which is disjoint from C	Axiom 3; naming
4. Let r be a member of D	Definition of "committee"; naming
5. r is not a member of C	Definition of "disjoint"
6. There is a committee E of which p and r are members	Axiom 2; naming
7. C and E are not equal	Definition of "equal"; 5 and 6
8. p is a member both of C and E	2 and 6
9. p is a member of at least two committees	7 and 8
10. Therefore every person is a member of at least two committees	Generalization

Q.E.D.

This figure is reproduced, with a few changes, from an excellent undergraduate textbook*. It is used there to illustrate the workings of axiomatic systems, in preparation for developing the theory of non–Euclidean geometry. It is comparable to the kind of proofs that are given in advanced works but is much less complex, and the individual steps are spelled out in much more detail. The box shows three axioms, which have to do with committees and their members, and one theorem: "Every person is a member of at least two committees." This theorem does indeed follow from the axioms. This may be seen with the help of the diagram on page 71. The point of the example is to

* *The Non–Euclidean Revolution,* by Richard Trudeau. © Birkhäuser Boston, Inc., 1986.

give a completely rigorous proof. This purportedly rigorous proof is presented in 10 steps, in the bottom half of the box.

Without disputing the conclusion, namely, that it follows from Axioms 1, 2, and 3 that every person is a member of at least two committees, let us examine the claim that the written material located between the symbol "Proof" and the symbol "Q.E.D." constitutes a proof. There is no formal definition of what an acceptable proof is. There is an informal idea that a proof is a sequence of statements written in an unambiguous and strictly formal language which proceeds from the axioms to the conclusion by means of allowed and formalized logical transformations.

As we read through the proof, we find that there is one step that is more troublesome than the others. This is Step 7. We pause there, and our mind has to grind a bit before going on. Why are C and E not equal? Spell out the reasons a bit more. They are not equal because r is a member of E by line (6) but not a member of C by line (5); therefore, by the definition of the equality of committees, C and E are not equal. This argument requires that we keep in the forefront of our mind three facts and then verify, mentally, that the situation implies non–equality. This conclusion is deduced from the definition, which speaks only of equality. Thus, in our mind, we have to juggle simultaneously a few more facts: what equality means and how we can proceed to get non–equality out of it. In order to make clear what is going on, the author attaches to his exposition a symbolic diagram which he says is not really a part of the proof.

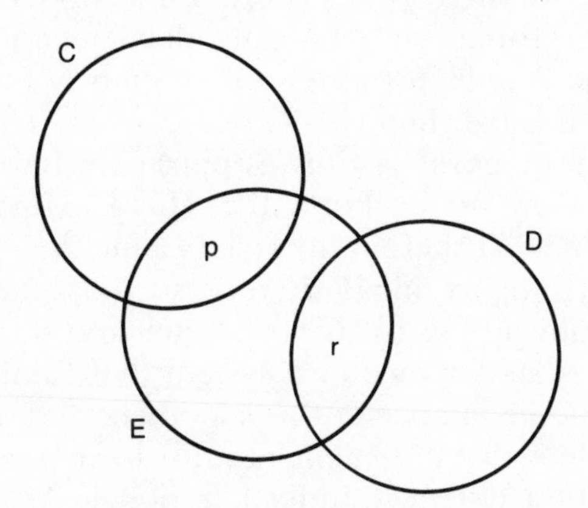

The picture (which is *not* part of the "proof") supplies the conviction and clarity which are not adequately achieved by the "real proof." This leaves us with a very peculiar situation: the proof does not convince; what *does* convince is not the proof.

In all human–human interfaces or human–machine interfaces, there is always the problem of verifying that what is asserted to be so is, in fact, so. For example, I assert that I have added two integers properly, or I assert that I have entered such and such data into the computer properly, or the computer asserts that it has carried out such and such a process properly. The passage from the assertion to the acceptance must proceed ultimately by extra–logical criteria.

This problem confronts us constantly. We find in the "reason" column of the above proof two mysterious words: "naming" and "generalization." There is no explanation of how "naming" and "generalization" are used in the proof. Now, if there is nothing worth discussing about these ideas in their application to the proof, why did the author bother mentioning them? Both are, in fact, difficult concepts, and philosophers have dedicated whole books to them. If they are important in the present context, how do we verify that the naming process or that the generalization process has been carried out properly?

Look at "generalization." In Step 1, a typical person is selected and named. Since it is a typical person, it is not specified which person it is. The idea is that if one reasons about a typical person, and uses only the characteristics which that person shares with all other persons, then one's deductions will apply to all persons (line 10). Should it not be verified, then, as part of the proof, that only those characteristics have been used? What are the formal criteria for so doing? By raising such questions, one can force the proof into deeper and deeper levels of justification. What stands in the reason column now, the single word "generalization," is pure rhetoric.

A rather different point is this. Suppose we have set up certain abstract axioms. How do we know that there exists a system which satisfies these axioms? If there is no such system, then we are not really talking about anything at all. If there is such a system, its existence might be made known to us by display: "such and such, with such and such definitions, is an instance of a system that fulfills Axioms 1–3." Would we then merely glance at this statement and agree with a nod of the head, or does the statement require formal verification that a supposed model of a system is, indeed, a model? Again we have been driven to a deeper level of verification.

The way out of these difficulties is to give up the needless and useless goal of total rigor or complete formalization. Instead, we recognize that mathematical argument is addressed to a human audience, which possesses a background knowledge enabling it to understand the intentions of the speaker or author.

In stating that mathematical argument is not mechanical or formal, we have also stated implicitly what it is—namely, a human interchange based on shared meanings, not all of which are verbal or formulaic.

Closure

Let us conclude by summarizing the import of this *mélange* of anecdote and analysis. The myth of totally rigorous, totally formalized mathematics is indeed a myth. Mathematics in real life is a form of social interaction where "proof" is a complex of the formal and the informal, of calculations and casual comments, of convincing argument and appeals to the imagination.

The competent professional knows what are the crucial points of his argument—the points where his audience should focus their skepticism. Those are the points where he will take care to supply sufficient detail. The rest of the proof will be abbreviated. This is not a matter of the author's laziness. On the contrary, to make a proof too detailed would be more damaging to its readability than to make it too brief. Complete mathematical proof does not mean reduction to a computer program. Complete proof simply means proof in sufficient detail to convince the intended audience—a group of professionals with training and mode of thought comparable to that of the author. Consequently, our confidence in the correctness of our results is not absolute, nor is it fundamentally different in kind from our confidence in our judgments of the physical reality of ordinary daily life.

Further Readings. See Bibliography

D. Berlinski (1976, 1986); W.W. Bledsoe and D.W. Loveland; N. de Bruijn; P.J. Davis and R. Hersh; N. Koblitz; S.K. Langer, vol. I, "Idols of the Laboratory"; L. Wos, et al.

The Criterion Makers: Mathematics and Social Policy

THE PIECE THAT FOLLOWS originated in a talk given in May, 1963, at the Annual Dinner of the Society of Sigma Xi, Auburn University, Auburn, Alabama. I recall travelling by train from Washington, D.C., to Auburn; shortly thereafter the station was abandoned. The country was already into the second generation of computers, and the examples I gave of computational mathematics applied at the consumer level were then actively being discussed. It is revealing to see which of these ideas has worked out and which has not.

At the end of each example I have placed in brackets a brief update.

Aside from the specific examples which are amusing enough, the main point of the article, I think, is still valid. Whenever mathematics is employed to recommend or to set in motion a certain course of action, it is done on the basis of a criterion. One never knows the validity of the criterion nor does one know the full implication of putting a particular criterion to work.

In computerized world economics, in computerized medicine, who sets the criteria and on what basis? As David Garfinkel has written:

> "We could wire up an intelligent artificial pancreas, but what would we tell it to do? . . . How much information should the patient be expected to communicate to an artificial pancreas, and how can the pancreas be made foolproof if the patient forgets to communicate with it or makes a mistake in doing so."[1]

Quis custodet custodies—who will watch the watchmen?

The Criterion Makers: Mathematics and Social Policy

The application of mathematics to the physical sciences and to technology is an old story. The theory of satellite orbits was worked on by mathematicians and astronomers for two hundred years following Newton. By the early 1900s, celestial mechanics, as it is called, was no longer of much interest. Though many difficult problems were left unsolved, the field attracted little attention. There were only a handful of people who would have known how to compute the orbit of a comet.

Today, the satellites have sent space theoreticians scurrying back to the work of the old masters. While John Glenn was making his flight, NASA's computing machines in Greenbelt, Maryland were grinding away to control and guide him, but the calculations they were performing could have been set up by the Marquis de Laplace, who lived in 1800.

A major impetus to applied mathematics has come in recent years from a new and surprising source: the social and economic behavior of human beings, people in the market place, people as they attempt to thread their way through masses of other people. As examples of recently developed disciplines within applied mathematics, we have game theory, operations research, statistical decision theory, data processing, information theory, econometrics, sociometrics, psychometrics, biometrics, human engineering, and others. These are all large and lusty babies, crying for attention and for the public bottle and all promising to grow up to become world shakers.

Though the abstract mathematical bases of these fields could have been promulgated many decades ago, there are good reasons why we had to wait till the 1950s to see their blossoming. In the first place, social situations usually require the handling of a great many pieces of interlocking information. Before 1946, when a man computed with pencil and paper, or even had an adding machine to help him out, ten simultaneous equations in ten unknowns would make him groan. Twenty equations would be at the limit of his endurance. Now social situations can involve hundreds, thousands, millions of people, and when the interplay is cast into mathematical language, billions and trillions of arithmetical operations may be involved in order to arrive at a prediction. Coping with arithmetic in this bulk requires a computer. Computers with great capacities for retaining information and with unbelievably fast operating speeds are commonplace, therefore there is now the possibility of making concrete the abstract formulations of econometrics, etc.

There is a second reason for the late arrival of mathematics in the social sciences, and it is a psychological one. For many years the social and economic scientists have suffered from an inferiority complex.

They look across the halls at their colleagues the physicists and envy their wonderful successes in planetary theory or in electromagnetic theory. "We must wait for our Newton," seemed to be their attitude.

The philosophers of physics took a more modest view of the achievements of their field. They pointed out that the successes were limited to relatively simple situations. Professor Philipp Frank of Harvard used to dramatize this point by holding a thin strip of paper in the air and allowing it to flutter to the floor. Though the basic laws of mechanics and aerodynamics are known, no physicist could predict the resting place of the paper. The philosophers pointed out also that since the collapse of classical physics at the turn of the century, the physicist has had a more modest idea of what physical law is. It now appears that such laws come not from the Cosmological Writ but from the minds of man; and that they are merely "models" of reality, of limited applicability, to be used or not used depending upon the purposes of the investigator.

From the soul–searchings of the physicist and from his notion of a theory as a mathematical model, the social scientist took heart. He crossed his fingers, he simplified, he built models, he hoped that on this platform of work a Newton might stand, but he would settle, in his own day, for a Galileo or even a Ptolemy.

But I must return to mathematics and social policy. It would be out of place for me to give formal definitions of these new areas of applied mathematics, but I must show how they can affect our day to day life. I will give a number of examples. Some are familiar: they are the stuff out of which misleading Sunday supplement articles are written. Some are for the future. All illustrate the way that the mathematical outlook can influence social policy. All illustrate the method, the attitude, the process of criterion formulation that makes the whole thing work.

A. *Marriage by UNIVAC.* "America Makes a Marriage Machine" says an alliterative front page headline of the *London Observer*, Sunday, February 25, 1962. There is available to a data–processing machine a large pool of information on men and women of marriageable age; their ages, social and religious backgrounds, occupations, education, interests, requirements in a mate. For a given person selected, the computer will go through its store of information, and come up with the names of those of the opposite sex who would make the best matches. What is crucial here is an *a priori* formulation of the measure of the "distance" or the "closeness of match" of a given male, female combination. This measure, which may in part be numerical and in part logical, is the criterion which the machine uses for selection. The machine simply computes this measure for all pertinent combinations of individuals, and selects those where the measure is best. The machine does not formulate the criterion. This is done by the programmer. The mathematics here consists of the notion of the measure, the notion of minimizing this measure, and the utilization of computers to carry out the bookkeeping rapidly and efficiently. It is, in fact, a modernization of the way in which marriages have been made in most societies in many periods of history.

The idea, at the moment, is a gimmick. It is shocking to those who prefer to see marriages made by the random processes of romance and to those who view a machine as a blind instrumentality rather than as the updated little black book of an efficient and knowledgeable matchmaker. It is an interesting development to those who believe that marriages ought to be made on the basis of rational selection. At any rate, there is no evidence to show that UNIVAC marriages are less successful than those contracted by couples who have met on the escalator at Macy's. Quite the contrary, according to my London article, the Scientific Introduction Service of New York claims just one divorce from 500 marriages made in this way since 1956. The national rate is one in four marriages.

[See the section that follows entitled "The Computerization of Love."]

B. *Occupational Placement.* The idea of testing to determine occupational placement is related to "A", but is more than a gimmick. Deep theories of mathematical statistics are employed in the formulation and validation of the tests. Millions of dollars and thousands of people are tied up in it. The success of the method varies from application to application. Testing for scholastic achievement, for instance, has shown considerable predictive value. On the other hand, although extensive work has been done in the area of pilot selection, for example, some critics feel that after one has accounted for obvious factors such as health and general intelligence, further placement testing has accomplished little. A criterion must be formulated of the best measure of fit between individuals and occupations.

One unpleasant aspect of reducing an individual to a few marks on a sheet of paper or to a few holes on a punch card is that his future can then hang by an irrelevant thread. Here is a true story I heard recently:

A student at a midwestern college was given a series of tests when he entered. On the basis of these tests, the dean advised him that he had a bright future ahead of him in architecture. The student took appropriate courses, did poorly, and was miserable. Two years later, the dean called him back and told him (ha, ha! and chin up, Smith!) that the office had confused some of the identification numbers.

[The computer today is doing an increasing amount of reservation and scheduling work (airlines, theatres) and there are companies that match skills with positions by computer. The admissions offices in the Ivy League Colleges still reserve the right to handcraft their selections. In some of the larger state universities admission or rejection is done automatically by computer on the basis of SATs, grade point averages, or other similar criteria. In some states, the legislature has prescribed the mathematical criteria for admission. In some schools, admissions people talk about the differences between a 2.43 person and a 2.3 person with every confidence that there is a difference between them. In general, it seems to be the case that the more selective a college is, the less likely it is to rely on a formula.

See also the sections that follow entitled "Testing" and "Mathematics as a Social Filter."]

C. College Placement. Every year the nation's medical schools produce thousands of interns who must be placed in hospitals all over the country. Each doctor has certain qualifications and certain preferences as to where he would like to go. Each hospital has a certain capacity and certain requirements. Until several years ago, the assignments were made by hand, using trial and error—a formidable job. Today, the assignments can be made by computer using the mathematical methods of linear programming. What is necessary here is to formulate a criterion for determining which, of two given total assignments, is the more desirable. The doctors are given their choices consistent with the internal constraints of the total assignment. Assignments have been produced by a machine and have been surveyed by a reviewing board. The results are said to be a great satisfaction to all concerned.

This method could have wider applications. Think of a college that has room for 1000 freshmen and has 5000 qualified applicants. Each applicant has particular qualities, and the deans have a picture of what the freshman class should look like. Surely there is room for a bit of mathematics here to relieve the paper juggling. Think, if you like, of a still larger problem. Most applicants apply to two or three colleges. This results in an enormous duplication of paper. Perhaps groups of colleges ought to pool their admissions work. With a computer or two and a workable criterion for admission in the presence of competition, the admission officers could push a button, and go back to the blackboard and lectern where their hearts really are.

D. *Flow of Automobile Traffic in a Tunnel.* A stream of traffic is flowing through a long tunnel. Maximum and minimum speeds are posted. There are inclines at either end. Each driver accelerates and decelerates according to his driving habits and the behavior of the car ahead of him. There is a distribution of reaction times and of rates of acceleration. What happens in the tunnel? How does the flow proceed? What does the spacing of the cars look like statistically? How often does a car plough into the one in front? If bottlenecks form how can they be eliminated? What is the maximum safe capacity of the tunnel in cars per hour?

These questions can be answered by computational techniques known as "simulation." An abstract model is set up of what are considered to be the crucial aspects of the physical situation. The model is then "run" on a computer. This is a form of mathematical experimentation. Though it does not involve the manipulation of physical models, it has many features in common with it. The experimenter observes what happens and interprets the results. He can then change the conditions and again see what happens. He can raise the maximum speed to 60 miles per hour or lower it to 10. He can introduce a drunken driver, i.e., one with a very slow reaction time, and see what happens. He can observe the over–all patterns. For instance, it is known that the flow of traffic exhibits certain of the characteristics of gases, with shock waves of cars developing a traveling wave back through the tunnel.

In experiments carried out at MIT for the Port of New York Authority, a program of admission to tunnels in "platoons" was developed which eliminated bottlenecks that appear when the car density becomes sufficiently high. This system also increased by 5 per cent the amount of traffic that the tunnel could carry.

The National Bureau of Standards has been simulating the flow of traffic over a portion of downtown Washington, D.C. The model is complex, the code is long, the results are tentative, but you see what kinds of questions are potentially answerable. Where shall we put the lights, how shall we stagger them, what should be the posted maximum speed, should we make this street or that street one way?

As people multiply to the point where they collide, react, and combine with one another *en masse*, what is observed bears strong resemblance to gases where myriads of molecules are in mutual play. The laws of history, yet to be discovered, may be to the patterns of Spengler and Toynbee as the partial differential equations of gas dynamics are to the ancient saw that "nature abhors a vacuum."

The method of modelling and simulation, brought about in part by studies in probability theory and in part by the existence of extremely rapid and versatile computation facilities, offers wide potentialities for application. Many groups ranging from government laboratories to university business schools use it. One of the great advantages of abstract experimentation is that one is able to change the assumptions quite easily, whereas to do so in real life may be impossible or undesirable. We cannot safely introduce a drunken driver into a real tunnel and observe his effect under a variety of traffic conditions.

Although it promises much, the method is still primitive. An abstract model, after all, can mimic only certain aspects of reality. In a given situation it may be hard to pick out those elements that are important. In addition to the mathematician and the statistician, the sociologist or the psychologist may be needed.

[Systems regulating traffic flow have been installed in a number of metropolitan highways, in Los Angeles and in Detroit, for example. The successes in this field are modest. A measure of the irrelevance of computers may be seen from the example of Boston, Massachusetts, a high–tech, think–tank city if ever there was one. On Friday afternoons, there is choking traffic on the Southeast Expressway and the tunnel to Logan International Airport. It could be alleviated by more tunnels or bridges, not by computers.]

E. *Relief for the Straphangers.* A task of major proportions has been facing a group of scientists at Columbia University. They are studying the daily flow of traffic on the subways and buses of New York City. Their object is to promote this flow. Perhaps the way to eliminate the terrible crushes is to stagger work hours. If so, how shall the staggering be scheduled in different parts of the city and in different industries? A man working in Brooklyn reaches his home in the Bronx an hour after leaving work. One industry has natural ties with another industry and should work the same hours. To determine a good policy, one

Traffic jam in Boston, Massachusetts. (*Photographer: Wade Harper.*)

must unravel a knot of changing conditions and mutually contradictory requirements. When proposals are finally made, the method of mathematical modelling and simulation will undoubtedly have played its part and the criterion maker his role.

[This is interesting because in the years that followed, the New York City subway system has been plagued by problems of a sort totally unanticipated by the Columbia University model makers.]

F*. Life and Love in a Small Town.* Still more complex is the economic life of the nation. Economists can argue night and day about the domestic and international effect of lowering the cost of autos $100, or of raising the minimum wage 25 cents. With modelling and simulation, they can study the effect of changing economic conditions. These techniques offer the possibility of converting economics into an experimental science.

Recently, at MIT, certain aspects of the economic behavior of a small town of 2000 people were modelled and simulated. The population was distributed appropriately in age and in economic status. Month by month, these abstract people, inhabiting the memory cells of an electronic computer, marry, have children, get a raise, lose their jobs, buy a house or a car, die, all on a probability basis according to reasonable statistics. The computer invents and keeps track of hundreds of thousands of transactions. A ten–year period has been simulated on an IBM 704 computer in 10 hours' time. The results are promising and it is not too much to expect that, in the future, modelling will yield information that will be of importance to the economist, the industrialist, and ultimately to the legislator.

[The system analysis of the "Life and Love in a Small Town" variety have been performed by numerous institutes, MIT and the Club of Rome among them. They have come in for much criticism, and few people, other than the systems developers, have any confidence in the meaningfulness of their predictions.]

G. *Peanut Butter and Potatoes.* What foods should be purchased and in what amounts to provide a life–sustaining diet at a minimum cost? This problem falls under the heading of "linear programming" and has been solved in special cases. For example, selecting from among 77 different foods and asking for minimum requirements of 9 nutritive elements, and using 1939 prices, the low cost diet was found to consist of flour, corn meal, evaporated milk, peanut butter, lard, beef liver, cabbage, potatoes, and spinach. To arrive at an answer, one must calculate the costs of thousands of possible diets. The computing machine, using the theorems of linear programming, can do this in a reasonable time.

For the undernourished nations, perhaps it is important to know at any given time of what the list consists and to solve the problem subject to the constraint of local availability and religious and social taboos.

[Diets in the advanced nations seem to be governed more by fads and quick reactions to medical findings than by the computer determination of idealized diets.

The mathematization of diets may develop a political aspect. In 1985, the National Academy of Sciences "drafted a report calling for lower recommended levels of some vitamins and minerals needed to maintain health. But some nutrition experts . . . expressed concern. They said the report . . . could be used by officials to justify further cuts in food stamps, school lunch subsidies and other feeding programs" (Providence, R.I., *Journal*, September 24, 1985).]

In a similar vein mathematically, but far removed from peanut butter and potatoes, is the question of the optimal allocation of resources. A company wants to put up a new plant. Where shall it build? Where labor is cheap? Where raw materials are plentiful? Where large amounts of power are available? Where the market is close? Mathematics can say much here, but someone must first formulate a criterion of desirability. Maximum profits to the stockholders? Maximum social benefit in some wide sense?

To make this dilemma real, consider the problem of New England. It has been losing its textile mills since 1920. The amount of human misery caused has been enormous. We can alleviate it either by introducing special conditions which maintain the mills there in contradiction to the social conditions which are forcing them out, or we can view it as a nation–wide problem on the optimal allocation of resources and, allowing the industries to fall where they may, alleviate the suffering by drawing on the total economic resources of the nation to relocate the affected population. The first solution turns legislators into special pleaders for their constituents. The second may be said to be the mathematical solution and the one which, despite obvious difficulties, must prevail in the long run. If this country dumps butter abroad, the man on the street in Denmark may be hurt. The computing machine offers the possibility of coping with the interlocking relationships of world trade. We are supposed to be our brother's keeper but are only gradually achieving the means to study out how one man affects his neighbor.

The examples ***A*** to ***G*** can be amplified many times. From the express counters in supermarkets to automatic diagnosis of disease to matters affecting defense policy, these are instances of the increasing importance of the mathematical approach.

The scientist, the computer expert, sits at his desk and smokes his pipe. Let us suppose he has a bit of the utopian in him. He dreams a dream. What is it? The world is complex. The population is increasing. The demands are many. The points of attraction between individuals and between groups are many. The antagonisms are many. Perhaps the power of mathematical thought and the use of computing machines will help us process the trillions of bits of relevant information. Perhaps we can better control the flow of people and the flow of goods and satisfactions. Perhaps here and there order can be brought out of chaos. Perhaps the hungry can be fed, crime and civil disturbance abated, wars quelled. Perhaps, in this way, we can be led to the gate of the Garden. With this thought, exhilarating but shocking, the dream evaporates and our dreamer awakes.

What are the dangers of the dream? I have said the dangers of the dream, but I really should have said the dangers of the method and the attitude—for the dream itself is a noble one.

1. The method can be misapplied, imperfectly applied, or applied to wrong ends. This possibility is so obvious—think of atomic power—that it needs no further comment except for us all to get up and shout it from the rooftops.
2. The method can become the basis of a scientific priesthood, most likely composed of second–raters, which isolates itself to protect its power and does not communicate with the public.
3. Worse still, as Norbert Wiener has pointed out, the machine, with the method built in, could become a sorcerer's apprentice, pouring out satisfactions at one end and chaos at the other, the satisfactions imparting to it the status of a holy, untouchable, un–turn–off–able object.
4. The method requires a criterion to work. The criterion must be extra–mathematical, extra–scientific. Conflicting criteria can cause strife and dissension. As the great comedian Willie Howard said years ago, "Comes the revolution, you'll *like* strawberries and cream." Today he might say, "Comes the mathematical revolution, you'll *like* peanut butter and potatoes."

There is a final danger that comes from the mathematical attitude itself.

As individuals, confronted with their own problems of daily living, mathematicians are as bright and as stupid, as liberal and as conservative, as nonmathematicians. Collectively, however, in their institutional capacity, so to speak, there is a certain attitude that emanates from their work. It comes from the great mathematical discussions of the past which are now being brought to bear on the affairs of the world. Mathematicians employ the axiomatic method with hypothesis leading to conclusion via logical operations. They are adept at symbolization, abbreviation, strict definition. They show a great interest in relations, in symmetry, pattern, and variation. They revel in abstraction and generalization. A mathematician is adept at distinguishing between what is arbitrary or conventional and what is basic.

The mathematical outlook displays great willingness to alter the hypotheses, to simplify and change the original problem in order to progress. The mathematician lives in horror of circular reasoning and of inconclusive proof. There is a universality to the method which appears to be beyond the clutches of dominant philosophies and politics. After all, we and the Russians and the Chinese agree on the multiplication table.

There is an opposite side to the coin. A curious neutralism emanates wherever mathematics comes into contact with the world of men. The mathematician approaches calmly many things that are emotion laden for others. The difference between monopoly and duopoly may be the basis of an antitrust suit lasting a decade. To a mathematical economist, it is "merely" an additional term in a set of equations.

Despite the potential applicability of his modes of thought to the real world, the mathematician exhibits an aloofness to it. He derives the statement p implies q and is interested only in the implication and its derivation, but neither in the antecedent p nor in the consequent q. "If you tell me what to maximize, I will try to tell you how to do it." This is what he says, and he means it. The onus of determining what ought to be maximized is on others.

The onus is not passed on so readily to others. The mathematician knows the method; the method demands a criterion to make it work. The mathematician will be forced, willy–nilly, into the role of the criterion maker. I fear he may be singularly unsuited for this role. He tends to be impatient. He tends to be rigid and authoritarian. He is a true–or–false man, a zero–or–one man, a right–or–wrong man, a yes–or–no man, but the world, we know, is more complex than this. It is full of paradoxes and contradictions, unresolved and unresolvable. Can the mathematical attitude resolve the paradox of Job? If it attempted to do so, it might do considerable damage.

When a man steps up to the bakery counter and takes a number, he is sped on his way, and that is good. But when he takes a number, he becomes, in part, a number, and there's the rub.

What is the solution? The solution, it seems to me, lies in the cultivation of strong values that lie outside science. We must have more education for scientists in the humanities, in history. We cannot afford to be ignorant technicians. We must have less rigidity of thought. We must avoid becoming a scientific priesthood. The solution lies in mixing science and technology with the rest of life in proper proportions. I preach the golden mean, an unexciting but vitally necessary doctrine.

We must remember that though Mathematics may be the Queen of the Sciences, Science is not the only principle of life. "The tree of life is larger than the tree of thought," said Immanuel Kant. I believe it. Let us cultivate men of the mind who will also be men of the heart.

Notes

1. David Garfinkel: "We could wire up . . .", Perspectives in Computing, Spring, 1984. IBM Corporation, Armonk, New York.

Further Readings. See Bibliography

P. J. Davis (1962); J. Forrester

The Computerization of Love

IN MANY CITIES of the United States there has existed for the last ten or twenty years something called "the singles scene." This is a social milieu of unmarried, usually divorced, men and women in their 20s, 30s, 40s, 50s, and even higher for whom the normal, traditional modes of couple–formation or match–making have broken down.

The existence of the "singles scene" in the U.S. is a consequence, first of all, of the high divorce rate; secondly, of geographic mobility, which means, especially in the high–growth urban areas of California, Florida, Texas, and Arizona, that many people are hundreds or thousands of miles from their parents, cousins, and childhood friends; thirdly, of a currently merchandized pleasure and action–oriented philosophy of life, or "life style" as it is called, according to which everybody ought to be sexually active as long as medically possible.

All this means that there are hordes of mateless persons ("singles"), under a strong social or psychological imperative to mate, whether for the night or for a longer period, and with no traditional social vehicle by which they might be mated.

In such a situation, in a profit driven, rapidly innovating society like the U.S., it could be foreseen that businesses would spring up attempting to fill this social need. The three most conspicuous types which have appeared are the singles bar, the personals column, and computer dating. To investigate the singles bar would require late evening fieldwork in a somewhat unscholarly environment, so we will limit our considerations to computer dating and the personals column.

First of all, some brief definitions.

Computer dating is a procedure by which you enter into a data bank your own description of yourself according to the computer's parameters. The computer wants to know your income, your height, whether you smoke, what academic level you reached in school, what kinds of music or recreation you prefer, etc. It does not ask whether you are patient or impatient, considerate or brutal, tolerant or bigoted, compulsive or "laid back." You also supply the computer with a description of your desired mate according to the same parameters. The computer searches through its data base and finds the names and phone numbers of potential matches for you and sends it to you, together with a bill, x dollars per name.

Much has been written about computer dating; when the first computer dating service appeared in Cambridge, Massachusetts, in February, 1965, it was a media sensation.

> "Within six weeks, some eight thousand local collegians had dated their computer choices. Over the next eight months, as word spread through the nation's campuses, more than ninety thousand students signed up. While the avalanche did little to clarify the mechanics involved, it established the term 'computer dating' as an American household phrase." —John Godwin, *The Mating Trade*, p. 80.

We do not intend to repeat this already over–reported story. We simply discuss this phenomenon as one more example of the mathematization of society. We should note, first of all, that computer dating need have nothing to do with computers. It can be done by hand, and indeed probably has been done by hand, as the following report indicates:

> "One inescapable conclusion that I reached after surveying the field was that a great number of companies don't use a computer at all. The computer crops up only in their advertising. I approached one large outfit in northern California with questions about the make of their particular machine. 'Make?' said the manager. 'Yeah, well it's down in Los Angeles.' 'Yes, but what make is it?' 'Oh, different kinds. We use different kinds.' And, after a moment's pondering: 'They're all very modern. We use only the latest models.'
>
> "Oddly enough, the vintage of the apparatus could not conceivably make any difference to the result obtained. When I asked computer executive Howard Rigi about this, he merely laughed. 'Given the usual number of participants,' he explained, 'the whole process is just a delusion. You cannot get specific matchings from programs based on multiple determinants and on a people–pool of, say, six or seven thousand. At minimum—I say minimum—several hundred thousand subjects would be required. And there can't be more than a couple of companies in the entire country who have that many.'
>
> "Yet the magic exuded by the word 'computer' is stronger than all the technical reservations of the experts. Especially since few, if any, of the computer clients have even the haziest notion of what precisely the machine does." —Godwin, p. 86.

The important thing is not the computer, or its absence, but the information–handling process in which the single supplies certain data which are ultimately transformed into a date. The information about each client is given by some single's filling out an answer sheet: "Yes or No." "Pick on a scale of one to ten." It doesn't matter whether the applicant actually fills out the form or whether an interviewer does it. The important thing is that the questions are chosen in advance, are the same for all candidates, and can be answered only in yes–no or numerical fashion. This results in a condensation or projection of a human being—the applicant or candidate—down to a few dozen check–marks on a sheet of paper (or corresponding bits on a magnetic tape or floppy disc). Such a condensation or digitizing is one version of mathematization. The set of checks or bits serves as a very crude mathematical model of the applicant, or of his or her needs and desires.

By way of comparison, here is a surprising little item from the Soviet newspaper IZVESTIYA, December 15, 1985, kindly brought to my attention and translated by Professor Ralph DeMarr:

> "There is such a service ________
>
> A BRIDE FROM . . . A FILE CARD
>
> "In 90 cases out of 100, those who apply to this matching service in ODESSA (on the Black Sea) find a mate.
>
> "Those who use this service first fill out a sociometric card and then consult with a psychologist. All information is then used to find a suitable mate. This takes about 3 months. Maximum age difference is about 10 years for matching partners.
>
> "Information cards are filed by number and always have a photograph. If the information thus filed is of interest to both clients, then a meeting is arranged." [Etc., etc.]

Two points are noteworthy: one, the singles scene, in some form, evidently exists under Socialism, at least in Odessa. Two, a filing cabinet in Odessa evidently can do much better (90 out of 100!) than a computer in L.A. or New York.

The traditional matchmaker was part of a traditional community. That community (whether Jewish or Italian, Polish or Chinese) supplied a context within which judgments could be made as to who was a suitable mate for whom, by traditional standards of age, wealth, and family status.

Such "quantifiable" parameters were by no means the matchmakers' only considerations. "Intuition"—a knack of divining the right match—was the key element in a matchmaker's success. This is something that can hardly be taught. It seems that a few people were born matchmakers, while the rest were merely grateful to them for their gifts.

When we compare it with the traditional matchmaker, we recognize the dating computer as a crude and incompetent attempt at an "expert system." This is an expression much used recently in the artificial intelligence business. "Expert systems" in general are computer programs that attempt to reproduce the best existing expertise in such fields as neurological diagnosis or specialized branches of organic chemistry. The dating computer does not compare in expertise with the traditional matchmaker. From the available statistics, it seems that the dating computers' success rate is far inferior to that of the traditional matchmaker, yet the traditional matchmaker is on the way out, and the dating computer is on the way in, as traditional communities evaporate and isolated singles proliferate.

What about the personals column? How does that fit into our perspective on the mathematization of mating?

The personals column is much the same as computer dating, except that the computer is eliminated. There are publications like *Singles Scene*, an Albuquerque, N.M., monthly, whose *raison d'être* is the personals column. Many well–respected publications serving a general audience also find it worthwhile to publish a personals column, like *Harvard*, a publication addressed to alumni and alumnae of Harvard and Radcliffe. The *Saturday Review of Literature*, a recently deceased fortnightly book–review magazine, was known for its personals column back in the forties when I was an undergraduate.

The personals column is divided into two parts, Male and Female. It is a specialized kind of classified advertising, in which you are selling yourself. The ad must do two jobs. First of all, it must communicate some minimal information to classify you. To save time and money, certain abbreviations are used. One advertiser calls himself a GWM: "gay" or homosexual white male. This information will be accepted by the *Berkeley Barb* or the *Village Voice*, but not by the Albuquerque *Singles Scene*. Another advertiser is a DJF (divorced Jewish female). One more important initial is S, meaning "straight," or heterosexual.

The second part of the personal ad seems to leave the realm of statistics and enter the poetical. Here the advertiser seeks to portray him or herself and his or her desired one in such a way as to attract as many letters as possible from desirable correspondents—and at the

Есть такая служба

НЕВЕСТА ИЗ... КАРТОТЕКИ

В 90 случаях из 100 находят спутников жизни люди, обратившиеся в службу знакомств, созданную недавно в Одессе. Она коренным образом отличается от всех остальных в других городах. Поисковая система здесь новая и, как показывает практика, очень результативная. Без объявления в газете, пользуясь специальной картотекой, абонент получает возможность выбрать наиболее оптимальный вариант предполагаемого спутника жизни.

Любой человек, прибегнувший к услугам службы, заполняет информационную и социометрическую карты, алфавитную карточку. После заполнения необходимых бумаг с абонентом проводит беседу квалифицированный специалист-психолог, который старается как можно лучше понять человека, находящегося перед ним.

Рассказывает консультант службы знакомств, ассистент кафедры психиатрии и медицинской психологии Одесского медицинского института Виталий Дмитриевич Высоцкий:

— Поисковая система дает возможность работникам службы быть не просто посредниками между людьми, а активно влиять на их судьбы. О своих абонентах нам известно практически все: их социальное и материальное положение, состояли ли они в браке ранее, имеют ли детей, хотят ли их иметь в последующем браке. Мы обращаем внимание на такие моменты, как предполагаемая зарплата будущего мужа (жены), специальность, жилье, национальность, возраст. Для нас важно знать круг их интересов: любимые книги, хобби...

Оптимальный срок для выбора кандидата и знакомства с ним — три месяца. Механизм же работы достаточно прост. Все абоненты делятся на возрастные группы, исходя из того, что разница в возрасте не должна превышать десяти лет. В картотеке карточки располагаются под номерами, но обязательно с фотографией и самой минимальной информацией о человеке. Впоследствии, когда карта отобрана, мы предлагаем более полную информацию, сопоставляем людей с учетом их индивидуальных запросов и пожеланий. В том случае, если партнеры, судя по анкете, подходят друг другу, мы устраиваем встречу.

Т. СТАРОБИНСКАЯ,
А. СТЕПАНОВ.

Marriage by index cards. From *Izvestiya*.

same time, perhaps, to repel letters from undesirable correspondents. We quote a few typical examples on page 92.

At first glance, one might think that the personals column is a refuge from mathematization; each advertiser has complete freedom to write his or her own script in as fanciful or poetic a manner as he wishes. But after reading a few dozen of them, one is impressed most by their repetitiousness. A few qualities are claimed and requested over and over. These desired qualities ("loves children," "likes outdoor sports") can be combined and permuted in only a certain number of ways. As

F–2116 43–year–old Sagittarian professional woman, active in peace movement, vegetarianism, metaphysics, holistic health, travel, outdoor fun (w/w rafting, flying, walking). Turned off by hunters, fishers, SMOKERS, and drunks. Seeking gentle, sensitive, affectionate male over 40 to share love, work, and play. Send address.

F–2134 Attractive SBF—25, mother of two, mature, strong, outgoing, computer technician, upwardly motivated. Seeks bright, caring individual who considers himself a professional, 25–35, of any race and isn't afraid to take the bulls by the horn, during the good and bad times. Desires a relationship based on communication, common goals and a lot of teamwork.

M–2332 I am Spanish/Anglo, 35, 5′8½″, 165 lbs., wavy brown hair, moustache, hazel eyes. No children, live alone, non-smoker, self–employed, open–minded, good humor. Interests in art, music, nature, hiking, "fishing," movies, good wine and new and different things. Do we have something in common? Photo and phone.

(*Courtesy of "Albuquerque Singles Scene*".)

a consequence of this fact, the personal ad turns out to be in essence only a slightly disguised and elaborated form of the dating computers' input format. One could easily go through a couple of issues of *Singles Scene*, compile a list of a few dozen qualities mentioned (either positively: "love mustaches" or negatively: "please no dope freaks") and then assign numbers, say 1 to 50, to these qualities. This done, each ad would be ready to encode numerically as an entry in the data base of a computer.

Thus we reach our conclusion: in the personals column, as in the dating computer, mating becomes an impersonal, mechanistic procedure. As such, it can be seen as a computerized or mathematized form of coupling.

This is not to say that love or sex in our society at large is in immediate danger of being mathematized or computerized. We only say that in a certain, non–negligible segment of our society this has already happened to some degree.

By no means do we imply that computer dating or the personals column are objectionable. On the contrary, they may help and are unlikely to do harm. Nevertheless, they do stand as sign posts of the changes that are taking place in our society, and as indications of the way even our most intimate needs may be digitized, quantified, mathematized.

Further Readings. See Bibliography.

J. Godwin

Testing

IN ORDER TO mathematize society — to convert it to a form where it can be processed by a digital computer — it is first of all necessary to mathematize society's principal components — people. To mathematize people means to encode them, or to represent them, by sequences of zeros and ones.

In many respects, we are already encoded. Our medical records are essentially lists of numbers: blood pressure and pulse at various times, blood counts (note the arithmetical turn of phrase) and so forth and so on. Our school records, or "transcripts" as they are called, are also lists of numbers, denoted by the special term of "grades." You realize that "letter" grades are also arithmetical. One need only decode from A,B,C,D,F to 1,2,3,4,5. And then there are our tax records at the IRS. Numerical indeed!

But these numbers don't suffice for the purpose of mathematizing society. The principal social function of each of us is not as a pupil or a patient. It is, as an employee or worker, to participate in the economy. So our places and positions in the economic system have to be mathematized or digitized.

How do we do this? We take a test! The nation's grand central test–maker, the Educational Testing Service of Princeton, N.J., sells tests to the CIA, the Defense Department, the National Security Council, the government of Trinidad and Tobago, the Institute for Nuclear Power Operations, the National Contact Lens Examiners, the International Council for Shopping Centers, the American Society of Heating, Refrigerating and Air–Conditioning Engineers, the Commission on Graduates of Foreign Nursing Schools, the Malaysian Ministry of Education, the National Board of Podiatry Examiners, and the Institute for the Advancement of Philosophy for Children. In some parts of the country you cannot become a golf pro, a real–estate salesman, a certified moving consultant, a certified auto mechanic, a merchant marine officer, a fireman, a travel agent, a certified business–form consultant, or, in Pennsylvania, a beautician or a barber, without passing an ETS test. (Lists copied from "1983: The Last Days of ETS," by David Owen, Harper's, May 1983.) And we have not even mentioned the great vortex and center of testing — the schools — from kindergarten through graduate school. The tests we take in K–12 are a major part of our work file, the raw material for the computation that will determine what kind of work we get to do and how much we get paid for doing it.

In June, 1985, an unexpected and tragic death occurred in Jakarta. The Indonesian Minister of Education and Culture, Professor Dr. Nugroko Notosusanto, suffered a brain hemorrhage, at the age of only 54 years. He was buried at the Heroes Cemetery, Kalibata, with full military honors. According to the English–language Jakarta press, Dr. Notosusanto brought about a major reform in Indonesian education. He established a uniform nation–wide test for completion of secondary school. Unfortunately, almost all the candidates failed the test. To correct this calamity, he decreed that the test results should be "curved" — that is, fitted to a normal bell curve. When this was done, the result was that almost everyone passed the test. This, it seems, was a second calamity. Dr. Notosusanto was quoted as saying, "It seems no matter what I do people are angry at me." (Rough translation from Indonesian.)

At any rate, he had been in good health so far as his friends and family knew. This double calamity of testing was the only source of stress in his life mentioned in the English–language Jakarta press. Dr. Notosusanto's most notable accomplishment (again according to the English–language Jakarta press) was an increased emphasis on religion (Islam) and history (Indonesian history as a tool in "nation–building"). He was regarded as a sensitive, conscientious man, very faithful to duty. In view of that description, it seems possible that he was actually killed by the stress due to the problem of testing.

It's worth giving a moment's thought to this sad story. It casts in sharp focus the issue of what testing is really all about.

In particular, it is *not* just a matter of examining the students (i.e., in the simple literal sense of the word, looking them over), deciding which are OK, by some clear cut, "objective" standard, and which are not OK and then labeling them as such. If that were the case, it would be inconceivable that the same set of high–school graduates taking the same exam would nearly all pass on one day and nearly all fail on another day. Rather, a test, as it is used nowadays in modern societies such as the U.S.A. or Indonesia, is a partition device, a method for sorting a population into two subpopulations. The first subpopulation will be admitted to some desired status. The second will be excluded.

Testing is also a way of putting people ("personnel") into computable form. Such and such a score means such and such a rank in the bureaucratic hierarchy. The decision can be *automatic* (capable of being carried out by machine) and *objective* (no human being appears openly to whose prejudices the decision can be attributed). Being automatic and objective, it may appear to its recipients or victims as something God–given; that is, inevitable, eternal, and unquestionable. In truth, of course, it is the opposite; it is temporal and questionable and avoidable.

There are two crucial questions about testing: is it accurate? and is it harmful? In other words, does it fulfill its intended purpose? and does it do harm in other respects? On the first point, the answer is that, when properly done, testing does carry out a certain task accurately. This task, however, is related to, but not identical to, that originally envisioned. Take intelligence testing, for example. Does an IQ test measure intelligence? It measures *something*; for it does have, within reasonable tolerance, the statistical attribute call "reliability." Reliability tells us that a statistical measure (i.e., IQ score) is meaningful. The question is, *what* meaning? Are we measuring intelligence or something else?

The IQ is exact. It is a number. Intelligence, on the other hand, is an amorphous, verbally–defined quality. How could they be *the same*? IQ purports to be in some sense an approximation or equivalent to intelligence. How could we justify or prove this claim for IQ? In order to do so, we would have to analyze "intelligence" into its various manifestations: ability to solve problems, to succeed in difficult situations, to behave in the most appropriate manner. But, what kind of problems or situations? A quick glance at an IQ test will show that the only problems and situations for which the IQ questions are pertinent are classroom or academic ones. The IQ test is really just an instrument for predicting success in school. (And far from infallible, even at that!) It has little bearing on occupational or professional success or success in business, in love, or in the other testing grounds of life. There is a dramatic meaning–shift in passing from the "intelligence" of ordinary language to the "I" of IQ.

While IQ is a reasonable predictor of success in school, it is of course not the *same* as success in school. Some people with a relatively low IQ do well in school, and some people with a high IQ do badly. What then, really, is IQ? The only honest answer is, it is what it is. More wordily, the IQ score is a measure of one's ability on the IQ test. If someone chooses to use it with another meaning than that, he does so at his (and his victims') risk.

The same argument applies to any other "objective" test of some "aptitude" or "ability." The test does not measure exactly the aptitude or ability we are interested in, but, rather, some artifact brought into being by the invention of the test. As a consequence, test–taking ability becomes a new and crucial "aptitude" for getting ahead in the world. This is no joke. The coaching for various tests (Medical School Aptitude Tests, Law School Aptitude Tests) has become a substantial industry in itself. In many medical schools, whose students must pass the nationwide "Board" exams after two years, the faculty supply their students with substantial, systematic coaching on exam–taking strategy. Would not those students' future patients be better served if the exam were cancelled and the extra time spent on pharmacology or anatomy?

This leads into our second issue: aside from its accuracy or inaccuracy in its intended measurement, does "objective testing" do any positive harm? On this issue tremendous controversies are raging. For instance, it is claimed that IQ tests and other "objective" tests are culturally biased. There is no question that results of these tests are being used to justify claims of inferiority of the non–white part of our population.

It is obvious that tests could be written which consistently demonstrate Black superiority. As it happens, no one is seriously interested in writing and administering pro–Black IQ tests, whereas the present tests, which give Blacks, on the average, slightly lower scores, continue to be used. One rationale for these results is to say that many Black people are educationally and culturally deprived, and the test merely reflects this deprivation. There is nothing wrong with the test, only with the reality it reports. Although this defense may seem plausible, it is fallacious insofar as it treats the IQ as an objective, God–given measurement; we know the IQ is man–made and arbitrary.

A more serious defect in this defense of IQ is that it treats IQ as a purely passive institution, something which merely reflects what is. On the contrary, the IQ also affects what is. IQ and other culturally biased tests are part of the apparatus which restricts and makes more difficult the attempts of many Blacks and other non–whites to rise in our socio–economic order. Naturally, these tests are attacked by the political organizations of Black people and their supporters.

For example, Arthur R. Jensen (*Straight Talk About Mental Tests*, The Free Press, New York, 1981) has gained international fame by reiter-

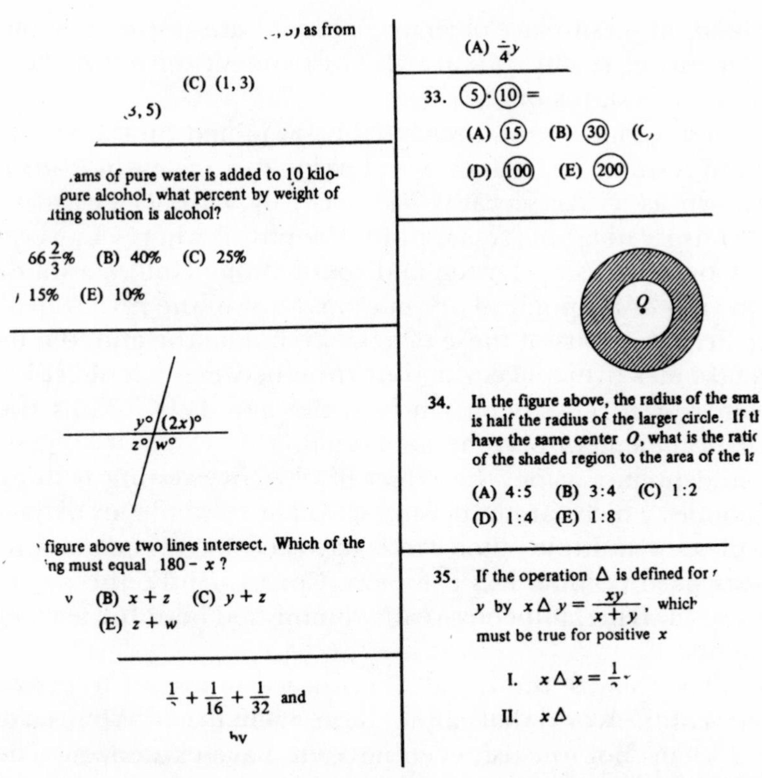

. . .) as from

(C) (1, 3)

, 5)

. . .ams of pure water is added to 10 kilo- . . . pure alcohol, what percent by weight of . . .lting solution is alcohol?

$66\frac{2}{3}\%$ (B) 40% (C) 25%

. . . 15% (E) 10%

. . . figure above two lines intersect. Which of the . . .ng must equal $180 - x$?

. . . (B) $x + z$ (C) $y + z$

(E) $z + w$

$\frac{1}{\cdot} + \frac{1}{16} + \frac{1}{32}$ and

(A) $\frac{\cdot}{4}y$

33. ⑤·⑩ =

(A) ⑮ (B) ㉚ (C . . .

(D) (100) (E) (200)

34. In the figure above, the radius of the sma. . . is half the radius of the larger circle. If t. . . have the same center O, what is the rati. . . of the shaded region to the area of the l. . .

(A) 4:5 (B) 3:4 (C) 1:2

(D) 1:4 (E) 1:8

35. If the operation Δ is defined for . . . y by $x \Delta y = \frac{xy}{x + y}$, which. . . must be true for positive x

I. $x \Delta x = \frac{1}{\cdot}$ . . .

II. $x \Delta$ . . .

Portions of questions from *Taking the SAT*. College Entrance Examination Board, 1985. (*Reprinted by permission of Educational Testing Service, the © holder.*)

ating the claim that the average difference of 15 points between IQs of Blacks and Whites is genetic in origin. A strong attack on Jensen's view is contained in *The Mismeasure of Man* by Stephen J. Gould, W.W. Norton and Company, New York, 1981. Related in spirit to Gould's is *The Science and Politics of IQ*, by Leon J. Kamin, Lawrence Erlbaum Assoc., Potomac, MD, 1974, which details the connection of IQ tests with racism and anti–semitism ever since they were first brought to the U.S.

Most readers will choose sides according to their political and philosophical preferences. There are many citizens who are quite comfortable with the belief that objective tests have demonstrated that some folks are just superior to other folks. This kind of thinking is sometimes called "conservative." On the other side, people with a belief in social betterment or racial equality (liberals, Lefts, or what you will) are more likely to be convinced by Kamin's and Gould's arguments that IQ testing (and other similar "objective" testing) are not merely objective measurements of reality, but are also instruments of social control for maintaining the status quo.

Since these remarks are, unavoidably, weighted on the liberal side, let us try to restore the balance by referring to a review in *Policy Review* (well–known as a "conservative" journal) by Michael Levin, a well–known "conservative" professor of philosophy. Levin reviews both Jensen and Gould, praising Jensen and condemning Gould, as one might expect in view of his political orientation. Most of the review deals with the specific arguments of these two books, but, at the end, the political animus becomes overt: "Peering out from between [Gould's] lines are our friends Marx and Lenin and . . . the new Left." Thus does the mathematical element become politicized.

Race and politics aside, the effect of objective testing is to devalue those qualities which cannot be so tested. For example, in high–school English classes, multiple–choice tests have become very common, while essay tests have become less common. Consequently, the importance of learning to write has been greatly diminished both for teachers and for students.

As we mathematize the world, we proceed to lose or to throw away those parts of the world that cannot be mathematized. What isn't mathematized seems not to exist, even never to have existed.

We should never forget that a stroll in the woods or a deep conversation with a new or old friend are beyond mathematics. And then, when we go back to our jobs, as administrators, teachers, or whatever, let us still remember that numbers are only the shadow, that life is the reality.

Further Readings. See Bibliography

P. Houts

Mathematics as a Social Filter

IN THE LIFE and work of the teacher of mathematics, there is a strange contradiction. He or she studied mathematics by choice. He loves to lose himself in this ideal world of clarity and precision. There is nothing he would like better than to invite others to join him there. For someone who loves math, teaching math should be a ball.

Sad to say, it isn't quite that way. Many of the students in mathematics classes are there by compulsion. Often they have little taste for mathematics, and many have great difficulty learning even a very little bit of it. What mathematics teacher can forget the first time he showed a class something especially elegant, something beyond the basic facts and problems in the text book? When the presentation is complete, a hand goes up. "Will we be responsible for that on the final?"

Certain complaints are heard in the corridor, year after year. The differential equations students are poorly prepared. They haven't learned what they should in calculus. The Calculus III students are poorly prepared. They haven't learned what they should in Calculus II. And so on back to Calculus I, and to College Algebra, all the way down to Intermediate Algebra.

We change textbooks, change curricula, but the complaints don't change.

The conclusion is to blame the high schools. Many college math teachers are convinced that most high school teachers do a terrible job. The college teachers whom I have heard express this view never visit the high schools and never talk to high school math teachers. The high school teachers I've met are intelligent and care sincerely how well their students do. The problem, they explain, is that the students coming into high school don't know what they need to in order to do high school work. The trouble is in the middle schools. And of course if you look into the middle school, you'll be told that the trouble is in the elementary school. In the elementary school you'll be told that the problem is in the home and the family. Thus everyone is to blame, and no one is to blame.

The situation I have just described from the vantage of a college math teacher is just the back side of the "crisis" often decried in headlines in the nation's better newspapers. "Crisis of scientific illiteracy," "rising flood of mediocrity," "falling far behind Japan and even the Soviet Union."

From the viewpoint of international educational competition, the situation is indeed a crisis and calls for immediate emergency measures involving non–trivial amounts of money. Strangely enough, however, the actions that are taken seem to be rather modest, and their effect on the situation in the schools perhaps will also turn out to be modest. I would like to reconsider the situation, again from a teacher's point of view, and see if there is another way of thinking about the problem. My conclusions may turn out to violate the calls of our educational competition with Japan and the Soviet Union. If so, I can only hope that this will not be judged seditious. What is worse, they may turn out to violate the perceived self interest of my profession. For this, I will surely be declared seditious. But let us see.

The first little question that would be asked by an innocent observer is this: "Why are so many people studying a subject for which they have, it seems so little interest, affection, or aptitude?"

The answer, of course, is simple: "It's required."

Yes, but required by what? By whom? For what reason?

The requirement of calculus and differential equations for engineers is beyond challenge. You can't get through many an advanced engineering course without knowing differential equations. The same applies to physics and chemistry. The nature of the material that must be learned in those fields makes a mathematics prerequisite inescapable. But what about the requirement of the University's Business School? I have occasionally asked business students whether they ever used the calculus they had learned in Math 180. "Never," they reply. I suppose I could go beyond students and ask working business persons, executives, entrepreneurs and so on if they ever use calculus, but the very idea seems ridiculous. Of course they don't.

Why is calculus required by the University's Business School? I could go over there and ask around, but it doesn't seem really necessary. I do know two relevant facts: (1) Most other business schools require calculus. (2) The business school has many more applicants than it is able to accept.

These two facts suggest two motivations for our university business school's math requirement:

1. Lack of a math requirement could mean lower prestige for the school; conversely, instituting a math requirement means higher prestige in the world of University Schools of Business Administration. (This is the world that is real to a professor of Business Administration, just as the world of mathematics as a profession is the real world in the eyes of a college math teacher.)
2. The requirement of passing calculus cuts down the number of applicants, making it easier to decide whom to admit every year.

As to point 2), a remarkably frank statement was made by John Kenneth Galbraith in his book *Economics, Peace, and Laughter*. Commenting on the models of mathematical economics, he says this:

> "Moreover, the models so constructed, though of no practical value, serve a useful academic function. The oldest problem in economic education is how to exclude the incompetent. The requirement that there be an ability to master difficult models, including ones for which mathematical competence is required, is a highly useful screening device."

Galbraith adds a dour footnote:

> "There can be no question, however, that prolonged commitment to mathematical exercises in economics can be damaging. It leads to the atrophy of judgement and intuition . . ."

Granting that these two reasons are in effect, a spokesman of a university school of business could still argue that the calculus requirement has educational validity. Perhaps in some of their courses some of the teachers sometimes may wish to take a derivative or to find an integral of something. It would remain a matter of dispute whether the calculus requirement is really necessary. I think it isn't, but I don't claim I can prove it. The point that interests me is point 2). Mathematics is serving as a filter, a way to sort out those who will and those who won't be allowed to get a business degree.

It is being used in the same way, of course, with respect to engineering students, but there the role of mathematics is more deeply entrenched and its usefulness is more apparent. In the case of business, we have a different story. There seems to be no necessity to make math a requirement. There *is* a practical necessity to make a selection among the students who want to go to the business school. The business professors decide to use math for that purpose. Is that OK? How should we (math teachers) feel about it? First of all, there is nothing inevitable about the choice of math as a filter. Some other filters that could be used, or that have been used are: family connections, political connections, income, ability in sports, personal charm, brutality and aggressiveness, trickiness and sneakiness, devotion to public welfare, etc. The first five have been relevant criteria in admission of students to U.S. institutions of higher learning; the last three are suggested, somewhat in jest, for particular relevance to a school of business. Each of the criteria can be taken in more than one way. Take family connections, for instance: in American schools it has sometimes been helpful to be the son or daughter of an alumnus or alumna. In England, at one time, it would have been helpful to have aristocratic family connections. In Mao's China, on the other hand, it was helpful to be from a proletarian family and harmful to have bourgeois parents.

It must be admitted that consideration of the alternatives to math makes the use of math as a filter seem more reasonable. Compared to the other possibilities, a math requirement has an appearance of impartiality, of objectivity, and of rationality. How much easier for the faculty of the school of business to require calculus and rely on the math department to do the grading and sorting than to impose some other criterion, such as the last on our list, and have to interview hundreds of students and rate them on this criterion.

But what price do mathematicians pay for the privilege of being a gatekeeper, an obstacle which must be overcome in the race to that good job, that professional career, to which young Miss and Mister America know they are entitled.

In order to evaluate the use of the mathematics filter, we should consider it from three points of view. That of the business school (or the economists, or any other group that is using mathematics as a criterion of admission), that of the mathematicians (who have to administer the courses and tests constituting the filter) and that of society as a whole, especially the candidates and applicants who are being filtered.

From the viewpoint of a college of business administration, I suppose the whole thing works tolerably well. The pool of applicants is cut down without any effort or expense on their part, and without any blame or resentment accruing to the business school. After all, flunking the math prerequisite is something that takes place outside of the business school.

Next, we have to evaluate the math filter from the point of view of the mathematicians. How should we feel about a university's school of business having a calculus requirement? There is an easy, almost automatic, answer to that question. We should feel great! More math requirements means more math students. More math students means more jobs for math teachers. Our math department gets one or two new positions. Great!

This is a normal and logical response. One might call it the material or the physical response (as opposed to the spiritual response).

But there is another response mathematicians could make, which I would call the principled response. It would go like this: We already have too many students in our classes who don't want to be there. To have classes which are interesting, we need to have students who are interested. As for the students who are in our classes only by compulsion, let them go free! We do not desire their dormant bodies while their minds are elsewhere!

Well, this spiritual or idealistic mathematical response will most likely bring a smile to the reader's face. It just isn't practical! But, practical or not, it is part of the felt response (spoken or unspoken) of the mathematicians to proposals for increasing math requirements. We really don't want to teach unwilling students in required courses. We want to teach interested students in voluntary courses. Even if there is no possibility of getting what we want, the way we feel affects the situation and must be taken into account. It follows then, that, as to the mathematics filter, the mathematicians are ambivalent.

Finally, there is the general population, society as a whole, especially the candidates or applicants who must be passed or rejected by the math filter.

One effect of the math filter is to introduce a serious bias against Blacks, Hispanics and Native Americans. This bias has, of course, been noticed by the advocates and spokespersons of these groups. The general response has been to institute campaigns to raise the mathematical competence of these groups and to improve the mathematical education available to them. I think that significant progress has been made in these endeavors. At the same time, the general situation in which the math filter excludes disproportionate numbers of women, Blacks, Hispanics and Native Americans continues to prevail and is unlikely to change much in the foreseeable future.

There is no doubt that whatever improvements in math education will be achieved as a response to the math filter are a benefit to society as a whole and to the young people in question. But why is the filter itself sacrosanct? Why not challenge its justification in terms of the specific school or job for which it is imposed? There is no law of nature,

God, or government that everybody must know the quadratic formula. Mathematics is interesting and important, but so are art, religion, literature, and many other things.

In a just, more rational world, mathematics would be used as a filter only for posts for which it is demonstrably required. Such a change would be gladly accepted by the mathematics professor. We do not really want to be gate keepers and agents of exclusion.

A "Marxian" Analysis of the Role of Computing in Organizations

THE DIALOGUE that follows is an edited version of a taped discussion between Charles Strauss (CS) and Philip J. Davis (PJD). Charles Strauss is a computer scientist whose more than twenty–five years of experience include teaching, banking, scientific computation, and computer graphics.

Several other dialogues between CS and PJD occur in the course of this work.

Charles M. Strauss.

A "Marxian" Analysis of the Role of Computing in Organizations

PJD: What kind of organizations do you have in mind?

CS: All organizations in which computers are introduced on a large scale, in which there is a large capital investment in them; where there are a number of people involved in simply keeping the computer system going. I'm thinking of the kind of computers where there are operators; at least two shifts of operators. I'm thinking of an organization that employs a site manager to coordinate the operators and systems programmers, applications programmers, and all that kind of stuff. I'm thinking of the typical big IBM or DEC mainframe computers, insurance companies, banks, other kinds of offices, scientific organizations, any place where the computer has really become an integral part of the operation. Not just an adjunct. Places where, if you took the computer away, those guys would be left bleeding.

PJD: In such an organization, how many people would be involved in computing one way or the other?

CS: As direct adjuncts of the computing facility, I'd say fifty to a hundred.

PJD: So it's a fairly substantial number of people we're talking about. We're not talking about one or two.

CS: No, no, no. That would be peripheral computation. I can't think of a small computing facility like that, where there are only one or two people, *in toto*, involved with the computer. I can't imagine that the position of the computer in an organization like that would be vital. I'm talking about where enough of the flow of the organizational structure is built in, around, and about the computer, that the computer really has taken on the aspect of the lungs or the heart or the spleen; let's say it's a vital organ.

PJD: In such a place is there a fairly rigid way the personnel is organized?

CS: There's always a rigid organizational chart, but job descriptions and the organizational chart really have very little to do with what actually gets done, and how it gets done.

PJD: Are there different levels of people that might have the title of programmer?

CS: Programmer has become generic. It's so generic in meaning, in fact, that it now has very little meaning. I've seen it used as a very derogatory term: "he's only a programmer, you ought to speak to the analyst."

PJD: What are analysts?

CS: They are people who have been programmers for two years. I was brought up that the word programmer encompassed it all. If you were a programmer, that was like saying you were an engineer. There are beginning engineers, and there are super engineers, but, you know, if you are an engineer, then engineer is what you are.

PJD: What do you mean by a "Marxian" analysis?

CS: It's the Marxian *aperçu* that a new means of production gives rise to a new class.

PJD: What is the class involved here?

CS: It's the class of computer people, specifically the programmers and analysts, but really anyone who has to do with computing. They look to the computer and the organization built around the computer for their rewards.

PJD: If you have a core of people devoted to the computer in one way or another, then presumably there is another part of the organization that is computer–free, in a sense?

CS: Well, let's say it's computer–antithetical.

PJD: Would you explain the relationship between the two?

CS: It's one of unmitigated hostility: still, they're as interdependent as the parts of the body.

PJD: Which is top dog?

CS: That's like asking is the stomach more important than the brain.

PJD: Where does top management stand?

CS: Top management has, in the past, fostered this hostility. It still does, as far as I can see. Remember, I came to this view not from hanging around industry or business, but from site visits to research labs. The problem was very plain there. In business, you figure that everyone's going to hate the computer center, because it's always screwing up, and of course they *are* always screwing up. One of the reasons, apart from their own incompetence, although that has some part to play in it, is that impossible demands, both mentally and organizationally, are made on them. They are servants, they're supposed to do what they're told and get out. However, you can't treat them simply like a service, like the janitorial service, or someone who comes in and takes dictation, or even someone who comes in and does your legal work. Okay, that's a better analogy. No, it isn't a better analogy, because a bank or insurance company might have some internal counsel, some lawyers that they employ 9 to 5. But usually they pay a firm to handle their legal work and a firm to handle their accounting work: all those specialized professional things. They just go out and hire the service and keep it outside. Those are old fashioned ways—in the sense that people have been doing it for a hundred years anyway. It's well accepted, and no one gets all uptight when some law firm takes in a couple hundred thousand or two million dollars in fees from a large insurance company each year.

If they treated the computing function and all the things that computers are doing for the business with the same *respect*, as the Mafia would say, it might be the beginning of a turnaround, but I don't think they ever could. And I don't think they ever will, because the lawyers and accountants don't have this new means of production. Those professions are still paper and pencil sort of things, it's a very linear culture. It is, in fact, the same culture as the business man has. He does things with ledger sheets, and paper and pencil. He does things with face to face talking. He does not do things with pushing around bits and bytes and tape and information and data banks, and all that kind of stuff.

PJD: New means of production gives rise to a new class. What you're saying is that not only do you have a new class, but you have a hostility between it and the old class?

CS: Oh yes, as much hostility as I see between the urban proletariat of the 1830s–1840s in England and not only the country gentlemen who you could figure would despise the terrible, unwashed, unruly city masses, but also the country peasantry. The peasantry despised the proletariat at least as much as the gentry did.

PJD: Well, what are some of the consequences of this class warfare?

CS: Physically, it's that computing facilities, ostensibly for reasons of security, are always way off in a corner of the building. They're never like the accounting department or the trust department, or any kind of ordinary department scattered hither and thither. And you can walk into a room and you see lots of people with desks in the middle; where the executives have their offices around the walls. You walk into any office building in the country and you see that.

PJD: You mean the computing facility is something to be hidden away, like plumbing or air conditioning.

CS: Plumbing—yes. And for much the same sort of reason. It's catching; it reminds people of what they would rather not be reminded of. It's not only unruly and ugly, it reminds them of the fact. I think, when it comes right down to it, plumbing is no more ugly *per se* than accounting.

PJD: Does this result in low status for the computer people?

CS: No, it doesn't because they have their own class structure within the larger computer society. But within the particular organization, all of them are given the cold shoulder. In the cafeteria they do not make a lot of friends among the other people in the organization.

PJD: So the profession thinks along professional lines, rather than across insurance lines, or whatever the particular business is in which they're hired.

CS: Yes, it's partly that they themselves would rather have it that way. Even if they did want to be friendly, if they did want to think that they were dedicated to the insurance business rather than the computing business, I think they'd find it very difficult. I can think of no instances where that kind of cross–cultural step has been made. It's like street corner society. To get out of the ghetto gang, you have to make a deliberate break with the gang. You don't both run with the gang and start hanging out with your betters.

PJD: Do they feel themselves underpaid?

CS: No, by and large not. They might feel that they should have been promoted to analyst, or senior analyst, or whatever kind of silly names they've made for the different grades. As you climb the programming ladder the big step actually is getting to manage people. That's the difference between commissioned officers and non–commissioned officers.

PJD: What are some of the other consequences of this hostility?

CS: It means that encapsulated within each of these businesses they've got something that's supposed to be working for them, and it should be working for them a whole lot more than it is. All you ever hear from the outside of the computing center is gripes, about how the computer guys never respond, and it takes them so long to do anything, and they're always asking for more money for equipment, and you never see any results. The computer people say the rest of the organization don't know their asses from their elbows. They ask for impossible things; they never give us specifications. They give us obsolete equipment; they never give us enough money to keep up with the latest technical developments, training and all that kind of stuff. Now these sound like ordinary gripes, and I've been hearing them for 20 years, which is as long as I've been hanging around computers, but I never really thought—as I do now—that there was a deeper meaning. I always thought that it was like the army—a certain amount of griping is necessary.

PJD: Years ago, people who started businesses were very much concerned with the nature of the product. Then, as the businesses aged, they were taken over; the presidencies fell either to accountants or lawyers, and there were harmful effects on the life of the business itself. Is it conceivable in your opinion that a person might rise from the computer ranks to be the president of a large business?

CS: I can think of only one, and that was a large business that was in fact a large college—when John Kemeny became president of Dartmouth. I think the jury is still out on whether that was a beneficial or deleterious effect.

PJD: Kemeny was certainly a man who was friendly to computers, although he wasn't himself in the computer business.

CS: Yes, that's what I mean. If he had been 20 years later, he certainly would have been in it.

PJD: You haven't heard of such a case within commerce?

CS: No. In fact in the software houses that I know, the very bright guys who got together at the beginning to start it all had their turn at being the great leader or the great president. Now in the company's maturity, 10 to 20 years later, it is the managerial types such as lawyers and accountants who are the presidents.

PJD: Do you think it would help matters at all if the top management had a better idea of what computers are all about?

CS: No, I think it's sociological or Marxian or something. The fundamental problem is not to know what computers are about; it's in devising organizational structures that will minimize the misdirected effort whenever the computer people turn inward to their machine. This is the sort of disfunction that occurs where, for example, lawyers give their loyalty to the legal profession rather than to the clients they're supposed to serve.

PJD: Would you say, on the other side, that it would help the computer people if they had a better picture of what business was about?

CS: Where are they going to get it? It seems possible, because there are seminars, and there is a certain amount of frozen–faced smiling across the gap, to find out what the business you're in is supposed to be doing, if you're in the computer end of things. And you could certainly go to seminars on computer this and computer that if you're an executive and you want to find out what those crazy guys down the hall are up to. But the organizational structure is not set up for it—in every place I've ever seen it's set up as I've described. There's an invisible hand guiding the sociology. The computer and the people who tend it are treated as a barely necessary nuisance—an immense nuisance—and the management people wish they could do without them and they wish they could get rid of them, but they can't. This is a management problem.

PJD: What's going to happen? Project forward 10 years. Is the class conflict going to get worse?

CS: Possibly not, what with distributed processing, and time sharing and do–it–yourself, problem–oriented languages. Now, I've heard this siren song for fifteen years, but I see a couple of signs that these new developments might be alleviating the situation. If only by making more people less twitchy about the magic that the computer does. It's like if only chauffeurs drove cars, we'd probably respect cars much more. In the same way, distributed processing and problem–oriented languages—a computer on every executive's desk, or at least within easy walking distance, a terminal that he can speak to in more or less his own language—will tend to alleviate things.

PJD: Will these problem–oriented languages be created outside the particular business?

CS: Yes. Usually what's done is semi–custom tailoring. Bespoke tailoring is much too expensive. A few years ago, Brooks Bros. closed their complete, from the ground up, bespoke tailoring outfit. But you can still go to lots of places, and for about half what it costs to get a complete suit made, with several fittings, get "semi–fitted." What the software vendors now do is sell you something like housing or modular housing. They build the shell and sell it to you with all kinds of hooks for you to hang your own bells and whistles on. So it's kind of a cross between shell housing and compiler compiler. In other words, you can make it look to the user like what he tells you he wants it to look like.

III

COGNITION AND COMPUTATION

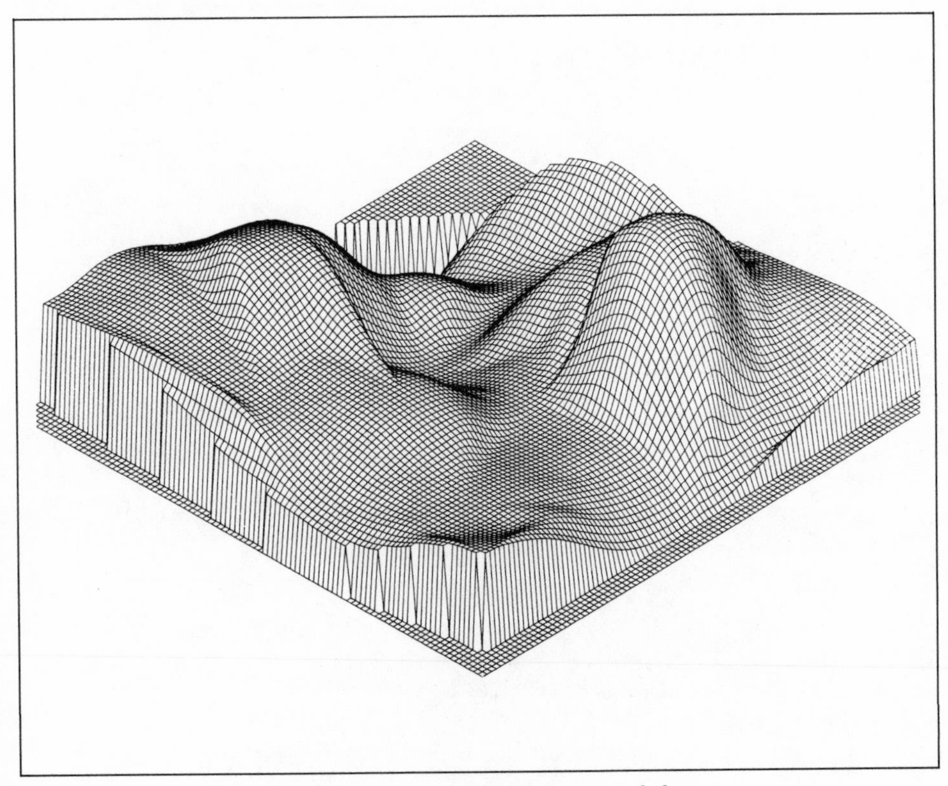

Interpolation to randomly scattered data.

The Descriptive, Predictive, and Prescriptive Functions of Applied Mathematics

WHEN MATHEMATICS is applied, it can serve to describe, to predict, or to prescribe. These modes are interrelated, but they are not identical. Let us elucidate their different natures.

Description. A hibiscus flower has five petals. The mathematical word "five" provides a certain description of the flower. This description serves to distinguish it from flowers with three, four, six . . . petals. It is not a complete description; on the basis only of the word "five," I cannot reconstruct the hibiscus flower very well. It does not tell me whether the petals are lenticular or circular. Amplify your statement to say that a hibiscus flower has five circle–like petals, and an image begins to form in my mind.

When a watch chain is suspended from its ends, it assumes very nearly the shape of the mathematical curve known as the *catenary*. The equation of the catenary is $y = \frac{a}{2}(e^{x/a} + e^{-x/a})$ (in appropriate coordinates and for an appropriate value of a). This equation provides an algebraic description of the shape of the suspended chain. With a bit of graphing, we obtain a geometrical curve from the equation, and then say to ourselves, "Why, yes, the curve does look like a suspended chain!" We may also reverse the process: photograph a chain, draw a coordinate system on the picture and then attempt to find an appropriate value of a so that the mathematical curve matches the photograph.

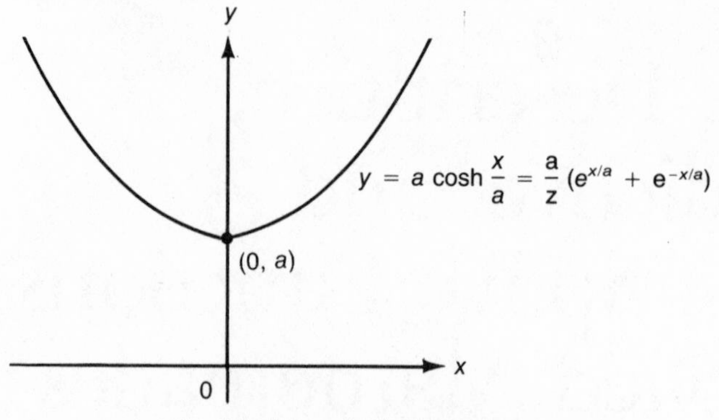

The Mathematical Catenary.

The orbit of a planet circulating about the sun is nearly an ellipse. In an appropriate coordinate system, the equation of the orbit is approximately

$$r = a\left(\frac{1}{1+b\cos\theta}\right)$$

The shape of a nautilus shell is very nearly an equi–angular spiral. The mathematical equation of such a spiral is $r = ae^{b\theta}$.

The "reverberation time" T for a concert hall is given by the Sabine–Eyring formula

$$T = -13.8\,\frac{\left(\frac{L}{v}\right)}{\ln(1-a)}$$

where L is the mean free path between successive reflections of sound waves, v is the velocity of sound, and a is the sound absorption coefficient. Though this formula was used for many years, it does not provide a very good description of what actually happens in a concert hall as determined by accurate electronic measurements, and has been replaced by a far more complicated computational procedure. (John R. Pierce, "The Science of Musical Sound," p. 146.)

In each of the above instances, something is going on in nature. By focusing on a sufficiently limited, sufficiently abstracted aspect of the phenomenon, we are able to summarize and replace that aspect by an equivalent mathematical description.

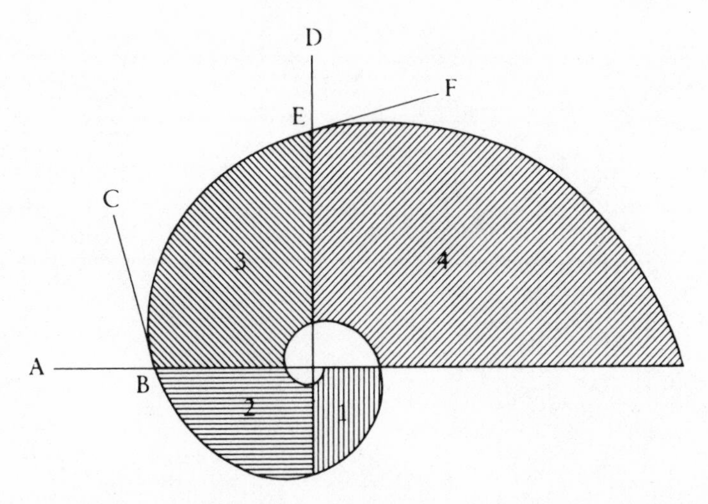

An equiangular spiral. The chambers marked 1, 2, 3, and 4 are exactly the same shape. *(From "On Size and Life," by T.A McMahon and J.T. Bonner, copyright Scientific American Books, 1983.)*

The chambered nautilus. Each of the chambers is geometrically similar.

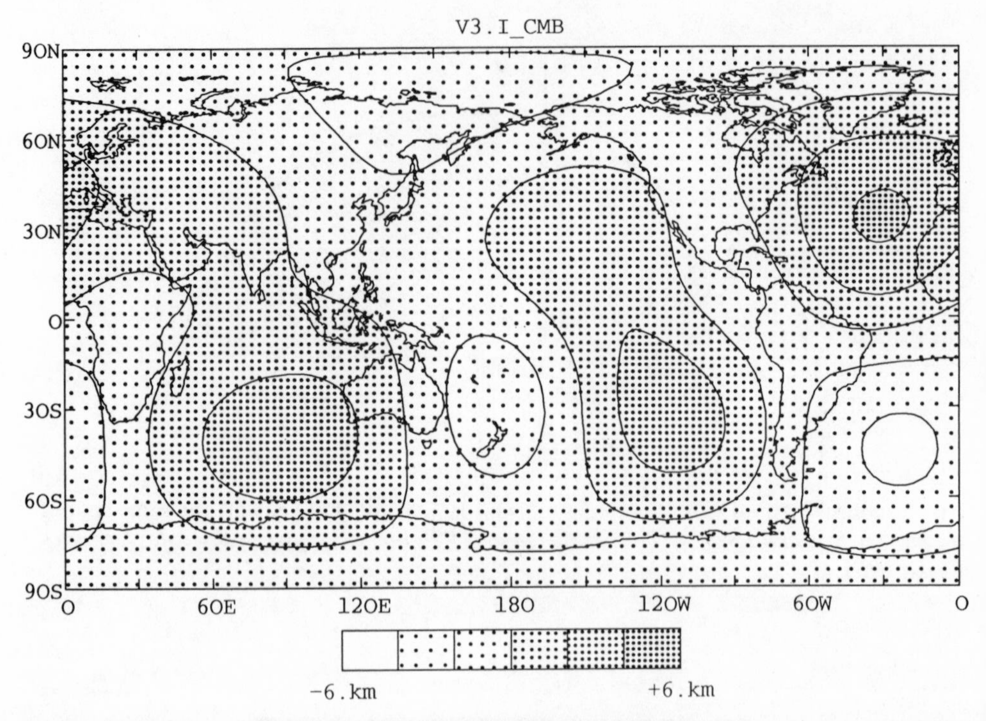

Seismology is the main tool in the study of the interior of the Earth. From the propagation of the elastic waves produced by earthquakes and recorded all over the world, a few important discontinuities are revealed. The core-mantle boundary marks the transition (at a depth of roughly 2890 kilometers) between the solid mantle and the liquid core. Its shape is not a perfect sphere, and we can nowadays map large scale "hills" and depressions of that surface. (*Courtesy of Andrea Morelli.*)

What aspects of the world can be described mathematically? We know many instances of successful descriptions, but, despite the extravagant claims that have been made for mathematics, there is really no convincing comprehensive answer to this question. Research in applied mathematics has as one of its functions the production of more and more mathematical descriptions.

What aspects of the world are expressible, but not amenable to mathematical description? In brief, I suppose they are the aspects that fall under the heading of humanistic statements. If a song–writer were to summarize life by saying, in the symbolism of the English language, that life is a bowl of cherries, then there is no mathematical equivalent to this. On the other hand, if a poet were to write that life is a game, meaning this in a metaphoric sense, then, since many games have been mathematized, someone might come along and provide a mathematical description of life which would have some claim to being an equivalent of the novelist's meaning.

There is also non–verbal, non–symbolic knowledge, for example, a feeling of well–being. There again, an experimentalist might say that well–being is simply a matter of hormone or blood sugar levels, and once again we see the possibility of description in mathematical language. Part of the conflict between scientists and humanists derives from a feeling on the part of the humanists that there ought to be a portion of the world that is immune from mathematization, and an opposing feeling of the scientists that all must be grist for the mathematical mill.

Prediction. The predictive function of mathematics is closely related to its descriptive function. Descriptions in the abstract symbolism of mathematics condense a great amount of information. The description may then be tapped or interrogated to answer specific questions. I hold my ten–inch watch chain by its ends. My fingers are at the same height and are four inches apart. How far below my fingers will the chain dip? On the basis of the known mathematical description (the catenary) I can predict this distance by a computation.

The word "predict," in normal usage, implies the passage of time. Predict, on the basis of Newton's Laws of motion, where Venus will be on January 1, 2001. Predict when and where the next total eclipse of the sun will be visible or what the weather in Fairbanks, Alaska will be two weeks from now. Predict how many multiplications you will need to perform if you attempt to solve a given mathematical problem using a given formula.

In the day–to–day world, prediction is for the most part carried out by non–mathematical means. If we are thoughtful voters, we should try to predict what will be the effect on us if Party *A* is voted into office. We do make such a prediction but not by formula. A person I like has made me an offer of marriage. I must decide: accept or not accept. I try to predict what life would be like under such an arrangement, but only the fatuous would accept the prediction of a point system devised by a mathematically besotted marriage counsellor.

The bases on which predictions are made are many: experience, reason, intuition, chance, "hunches," the advice of experts or oracles. When prediction is accomplished through the use of mathematical models and computation, it is called "rational" and is often granted a higher intellectual status then predictions arrived at by other means. The possibility of such predictions is part of the Cartesian world view. In the Cartesian frame of reference, to deny the possibility of "rational" prediction is to deny that the world can be understood.

Prescription. By the prescriptive function of mathematics I mean those situations where mathematics leads to human action or automatically to some sort of technological action. At the checkout counter, the zebra–striped cans are passed over the scanner, the bill is automatically recorded on the basis of the mathematics built into the system, and the customer pays. A flow of mathematics is set in motion and the outcome, which may be a number or an abstract symbolism, forms a prescription upon which an action is then executed. An oil drilling team is sent out on the basis of seismic records processed and interpreted mathematically. A course of action is prescribed in accordance with a medical scan. An astronomical expedition is outfitted on the basis of the mathematical prediction of a solar eclipse. A thermostat is set at 70°F, which then regulates the temperature of a house. We want to lay wall–to–wall carpeting in our study. We measure the study. It is eight feet by twelve feet. We go to the rug store and look at samples. We select one. The salesman informs us that that style comes in large rolls that are five feet wide. What to do? Further mathematics is required before a prescription is formulated as to how to take the shears to the rolled rug.

We are born into a world with so many instances of prescriptive mathematics in place that we are hardly aware of them, and, once they are pointed out, we can hardly imagine the world working without them. Our measurements of space and mass, our clocks and calendars, our plans for buildings and machines, our monetary system, are prescriptive mathematizations of great antiquity. To focus on more recent instances, and in this way to get a feeling for what living without them might imply, think of the income tax. This is an enormous mathematical structure superposed on an enormous pre–existing mathematical financial structure. In the United States, it has existed continuously only since the early years of this century, when a constitutional amendment was required to justify it. The effects of the income tax are felt by everyone. It gives rise to thousands upon thousands of jobs to provide advice, to interpret, and to administer its provisions. It affects social balances, financial strategies, computerizations. Do we

9/19/85 13:17:47.8 LAT= 18.18 LONG=-102.57 DEPTH= 33.0
MICHOACAN, MEXICO

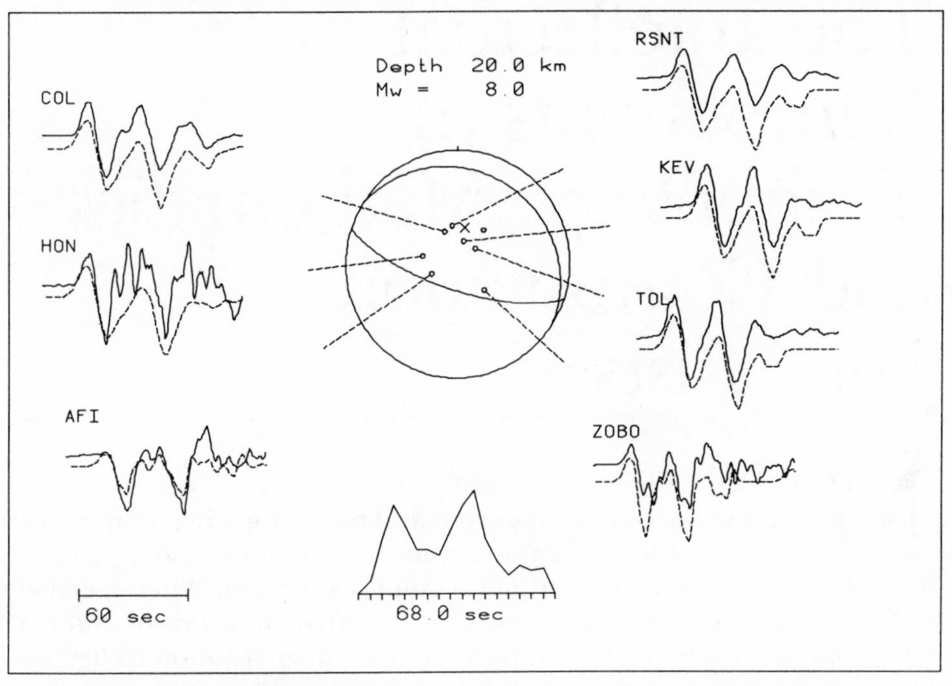

Mathematization of the great Mexican earthquake. (*Courtesy of Göran Ekström.*)

need this piece of prescriptive mathematics? It would be hard to cancel it; it is hard even to keep it at a constant level of complexity.

In American society, there are plentiful examples of recent or recently reinstated prescriptive mathematization: exam grades, IQ's, life insurance, taking a number in a bake shop, lotteries, traffic lights (Cairo, Egypt, doesn't have any), telephone switching systems, credit cards, zip codes, proportional representation voting, ration systems (in times of scarcity), point systems to qualify for mortgages, for civil service jobs, for driver license abrogations, for military demobilization, index systems for documents, school rating systems, the list goes on and on. We have prescribed these systems, often for reasons known only to a few; they regulate and alter our lives and characterize our civilization. They create a description before the pattern itself exists.

The Intellectual Components of Technology, Mathematics and Computation: Four Lists

MAKING LISTS, other than grocery lists or lists of the prime numbers, can be a dangerous and habit–forming occupation. After all, a list is only a summary, a condensation into a few lines of what may be an exceedingly complex situation. When it comes to lists that purport to explain one type of activity in terms of another type, we are on very shaky grounds. Some scholars caution us against making such lists, stating that such lists set up a meaningless question and then attempt to answer it. Can a few lines of language capture thoughts and activities that extend over hundreds of years and involve millions of people? Nonetheless, we like to make lists and read them.

Professor Wong Chu–Ming of the University of Rennes has made a list.

The Renaissance, the rise of science and technology, the subsequent industrial revolution, all burst upon the world in the West. Not in the East. This is a historical fact. There was plenty of technology and mathematics in the Orient; in many ways, the East was ahead of the West. But the explosion occurred in Europe. Why?

According to Professor Wong, the reasons why Chinese science lagged behind that of the West are:

"1. The rigidity of the social structure, making the class of literate functionaries the ruling class . . .
2. The fact that the written language was essentially literary and metaphorical in nature. It was very different from vulgar parlance and technical language, and thus inventors were able to make their discoveries known only with difficulty.

3. The contempt for manual labor symbolized by the long fingernails of the scholar; the contempt for mercantile activity which made the merchants one of the lowest social classes.
4. The logic, product of the essentially concrete and non–analytic language, which did not make possible the elaboration of axiomatic and deductive reasoning necessary for the development of science. The practical applications of the problem were more important than theoretical exposition.
5. The immensity of the empire, the wars and the revolutions, which explain why scientific liaison was difficult, and why certain important advances were lost or ill–transmitted . . .
6. In the West, man is the measure of all things, whereas in China every problem presented itself in a new light . . ."—Wong Chu–Ming, in *Scientific Change*, A.C. Crombie, ed., pp. 166–167.

This list appears as a summary comment to a long article by Professor Joseph Needham entitled "Poverties and Triumphs of the Chinese Scientific Tradition." Needham, who is a renowned authority on Chinese scientific history, commenting on Wong's list, allows that there is something in it. When you read Needham's commentary, however, you find that while all of Wong's points are mentioned, they are hedged in with so many "ifs," "buts," "maybes," and "on the other hands" that one begins to doubt the meaningfulness of the list. Still, there it is, and surely something can be learned from it. For example, it does not say that the technological revolution burst in Europe because the English and the Spanish brought the potato from America to Europe in the middle of the sixteenth century.

Let us take a look at a second list somewhat related to the first. In a thought–provoking book, F. Rapp tries to identify those requirements of the mind, those intellectual outlooks, which are essential for a technological civilization. Rapp comes up with eight. They are

1. Valuation of Work
2. Efficient Management
3. Impulse for Technological Creativity
4. Rational Thought
5. Objectification of Nature
6. Mechanistic View of Nature
7. Experimental Investigations
8. Creation of Mathematical Models
   a. qualitative exactness
   b. functional dependency
   c. abstract variables

In all fairness to this author (as well as to Needham and Wong), I admit I have lifted the points naked and unadorned from writings where they are clothed in considerable discussion.

When confronted with such a list, what I like to do with it is imagine what it would have been like if the items on the list were not present. Imagine what our country would be like if most of us didn't like to work, if our management were totally bogged down, if our thoughts were wholly superstitious or mystic, if we relied on the Great Authorities instead of looking around for ourselves and experimenting (or is our country already moving in these directions?). What would daily life be like if we thought of the universe as a holistic world–egg, if every phenomenon were related to every other phenomenon according to some hierarchical order, if our idea of physical processes were that of Zeus hurling thunderbolts into fields of misbehaving satyrs.

Ultimately, Rapp's list led me to ask a similar question for mathematics. What are the intellectual components of mathematization?

I reflected that the mathematical mind constantly turns over and examines its raw materials in accordance with its favorite operating modes. It likes quantity, pattern, symmetry. It likes abstractions. It likes precise statements expressed in precise, atomic, analytic languages that employ agreed–upon symbols and succinct abbreviations. It likes puzzles and their solutions. It likes deduction. It thrives on transformations, on standardized recipes, on "moves," logical or structural. It thinks in terms of yes and no, of true and false. Indeed, following Leibnitz, it actually asserts that one can build up the whole intellectual universe from zeros and ones. It believes, with Parmenides, that what is logical must be true.

The mathematical mind, confronted with a fuzzy, indeterminate universe, tries to find precise statements about that which is chaotic or random. It can see things probabilistically or statistically.

Out of such considerations I made my list.

Intellectual Components of Mathematization.

1. Ability to symbolize, abstract, and generalize the primary experiences of counting and spatial movement. A sharpened sense of quantity, space and time.
2. Ability to dichotomize sharply: yes, no; true, false; 0, 1.

3. Ability to discern primitive causal chains: If A, then B. Ability to concatenate and reason about such chains.
4. Ability and willingness to extract out of the real an abstract surrogate; correspondingly, the willingness to accept formal manipulations of the abstract surrogate as an adequate representation of the behavior of the real.
5. Ability and desire to manipulate and play with symbols even in the absence of concrete referents, thus creating an imaginary world which transcends the concrete.

This list isolates certain kinds of thinking, taking them out of the social and physical context in which they operate. It ignores our interest in the natural rhythms of life and of the world, and the strange physical phenomena that exist within it. It neglects the role of navigation and travel, of planning, of construction and destruction, of machinery and communication. It ignores, as sources of mathematization, trade and property and the legal and psychological props that relate abstract money to concrete things. It ignores metaphysical and religious imagery. Nonetheless, it is a list, and is open for contemplation and discussion within its limitations.

Let me push on. Is it too early to discuss the intellectual ingredients of computer science? Perhaps, but I will try.

Born of a mathematical mother and a technological father, computer science is still sufficiently close to the womb that the prototypical scientist in that field shares some of the traits of the mathematicians. The umbilical cord is now cut, and, as the subject moves farther away from its swaddling clothes, it becomes clear that a new type of professional is evolving.

Dedicated computer scientists prefer machines to people. They snuggle up to their terminals and clasp them in a lover's embrace*. They prefer action to thought. ("Hey, let me show you this great thing I've got"; not "Let me tell you about this great idea.") Nonetheless, when they are in a state of mental illumination, they often glitch the action. Thus, if a computer scientist has thought through the idea of a program, he is inclined to think that the program already stands implemented, bug–free.

* Cf. the title: "Cohabiting With Computers," J.F. Traub, ed., a collection of essays toward understanding the "meaning, promise, and challenges of the computerization of human society."

Computer scientists are impatient, brash, promising the world. They see Gordian knots all over the landscape and believe they can cut through them all by reducing the world to programs. They are especially impatient with those processes, such as politics or novels, that resist computerization. They tend to avoid such things and if, nevertheless, they happen upon something of that sort, they think: how very stupid, how irrelevant, how primitive it all is, but with a few decent algorithms we could fix it all up! Computer scientists are ahistoric; there is a present and a future, but absolutely no past.

They tend to force on a client what they are able to do, rather than thinking through what would be good for a client.

What, then, are the intellectual components of computer science? All the components of mathematics are, of course, relevant. They must, however, be refocussed. Here is my list:

1. Algorithmic thinking. Programs, recipes. The standardization of a particular craft through an analysis of its tasks into a sequence of easier subtasks. The frantic desire to see everything as broken down into the elementary decision of "yes" or "no".
2. Modular thinking. The formation of useful algorithms and their positioning as black boxes or modules where they are to operate as a standardized part.

In a sense, 2) is the reverse of 1). Whereas 1) analyzes or breaks tasks down, 2) synthesizes or builds them up. Here microdecisions are parlayed into macroeffects.

3. Systems thinking. This is an intensified version of 2), the assembling of a very large number of a few standardized parts into a larger structure. The organization of the components may be more important than the components themselves.
4. State thinking. Modality. This is exemplified by, say, the word processor programs: "We are now in the text–creating mode. We are now in the text–altering mode."
5. Metathinking. The creation of programs whose input is a program and whose output is a program.

All of these drives are now played out against a backdrop of psycho–social–technological circumstances.

We must have the willingness to go into certain areas formerly reserved for human activity and allow them to be replaced by a computerized surrogate. We must have the ability through appropriate

THE

INTELLECTUAL COMPONENTS

OF

COMPUTATION

MADE VIVID

BY

THE OPERATION OF

A

CHINESE RESTAURANT

(The details in the figures that follow are intended only to be schematic.)

Algorithmic Thinking

The process of making chop suey is algorithmized by a recipe and executed by a chef.

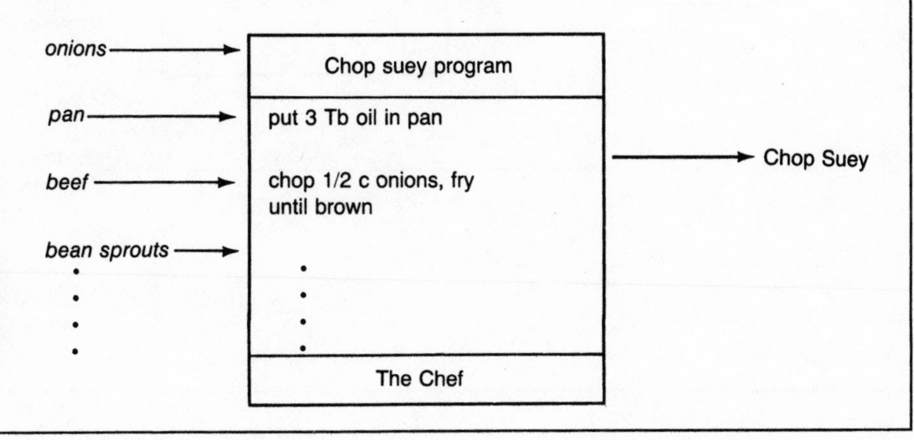

Modular Thinking

Various dishes have been algorithmized and interlinked as regards common inputs and common internal processes.

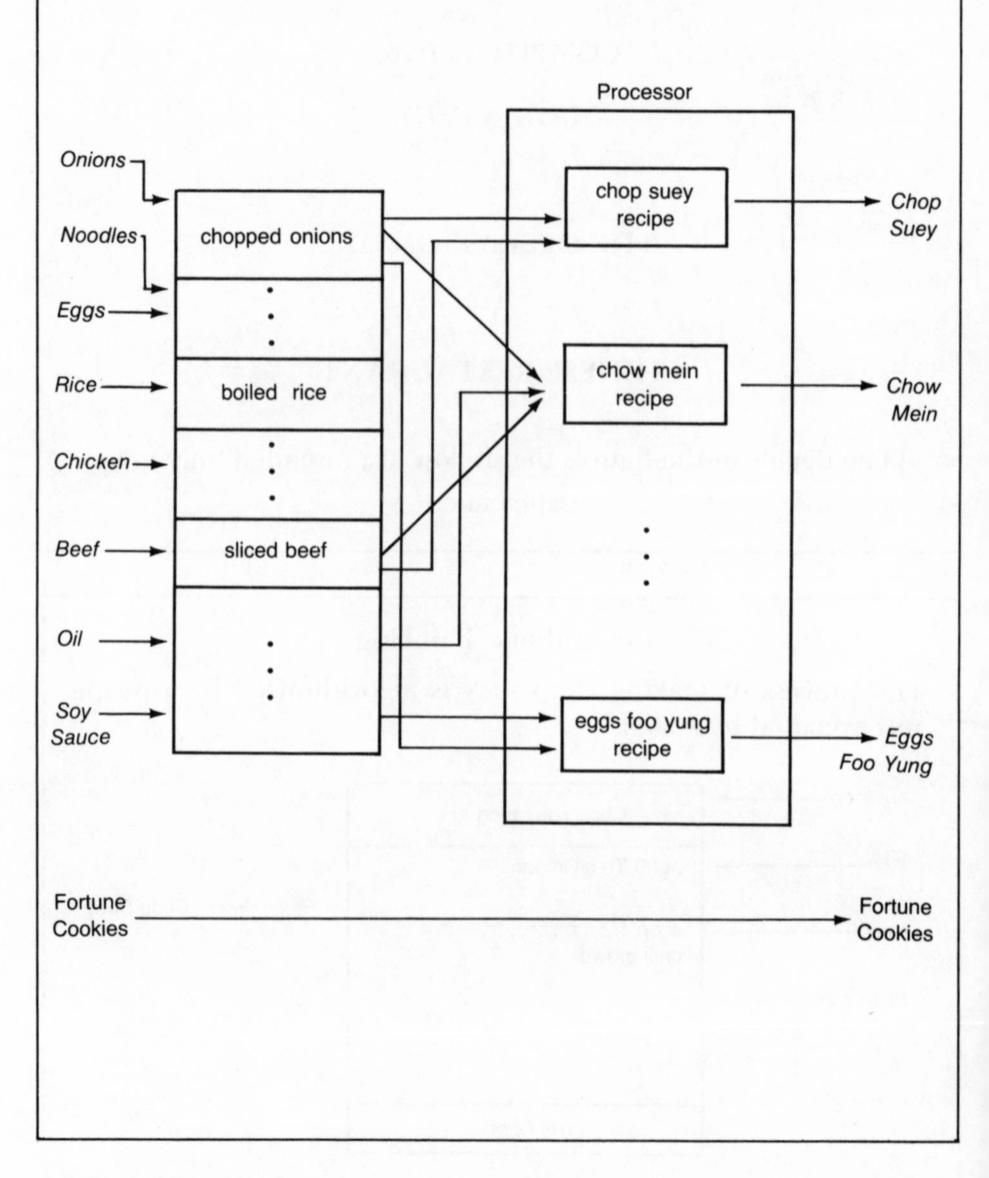

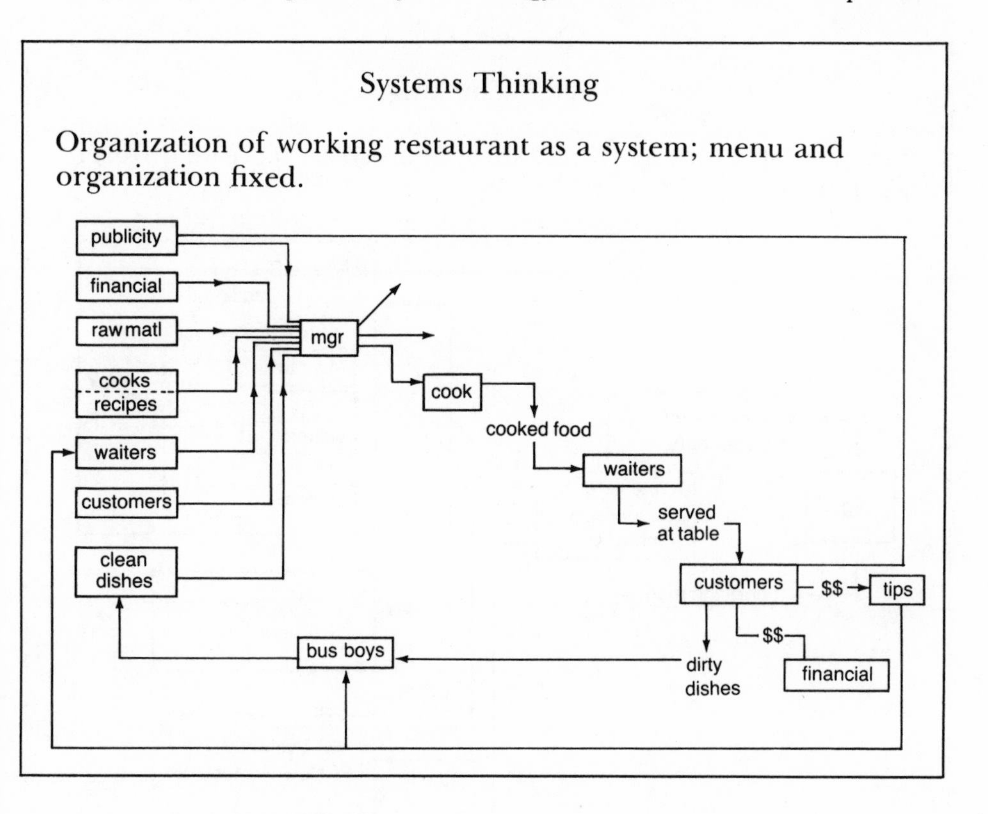

hardware to effectuate our abstract programs. Reciprocally, we must allow the hardware to suggest new forms of mathematical developments. In the past, the development both of abstract computer science and of real–world computer science has positively lived off the rising expectations of accomplishment in the field of computer hardware. It is interesting to contemplate where the field will be if the hardware reaches a plateau of development as did, for example, the steam engine.

If the popularity and the central position that autos occupy in our life can be explained on the basis of power and speed placed at the tips of everyone's toes, then the psychic involvement with computerization—whether it be of programmers, hobbyists, hackers—comes from the illusion of mental power placed at one's fingertips.

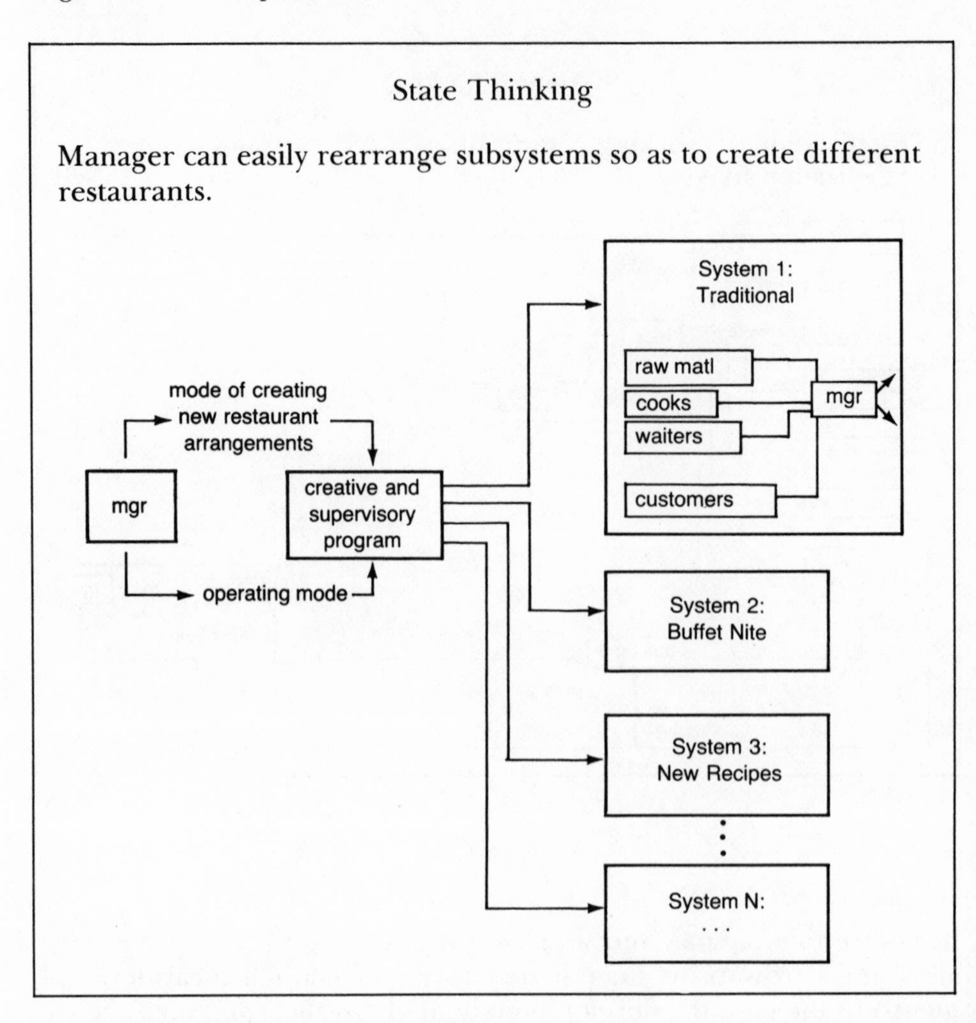
State Thinking
Manager can easily rearrange subsystems so as to create different restaurants.
mode of creating new restaurant arrangements
mgr
creative and supervisory program
operating mode
System 1: Traditional
raw matl
cooks
waiters
customers
mgr
System 2: Buffet Nite
System 3: New Recipes
System N:
. . .

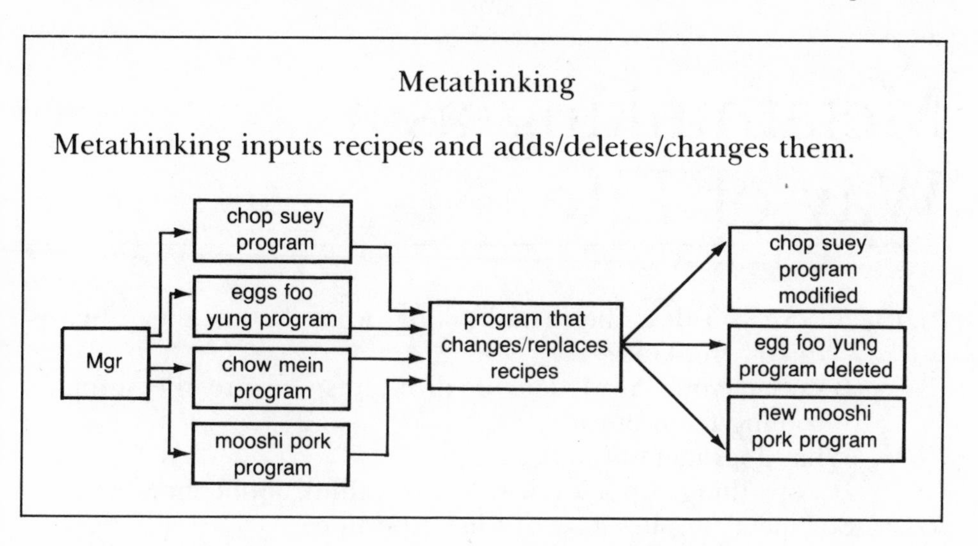

The prototypical and parochial image of the mathematical or logical genius is that of a big–brained, gnome–like, myopic, preoccupied, otherworldly individual. In single physical combat with one of the barbarians of the world, he would get trodden into the dust. For him, the intellect, and the power created by it, is the key to personal survival.

Further Readings. See Bibliography

S. Card, T. Moran and A. Newell; A. Crombie; J. Needham; F. Rapp; J. Traub; S. Turkle

Metathinking as a Way of Life

PJD: Before you describe metathinking, what do you mean by ordinary thinking?

CS: It's what your head does in direct response to the stimulus coming from outside.

PJD: What does metathinking do?

CS: It pops things up a level. It lets you think about thinking.

PJD: So "meta" implies a sort of introspection?

CS: It implies one level up, or an upper level looking down at the previous lower level.

PJD: Where does this occur within computer science? Is it within programming or within theoretical computer science?

CS: It is all–pervasive. It is natural to the way the original computers were built. It is the way they do loops and indexing and the fundamental techniques of pointer chasing. The fundamental technique of doodling around data and programs is by the program actually modifying itself.

PJD: Let's be very specific—let's get some examples of metathinking. You were saying that a program that modifies itself is an instance of this.

CS: Yes. The program treats itself as data. Ordinarily when you think of a program running in a computer, you think of it doing something with data somewhere else in the computer. However, the program itself can treat itself as its own data.

PJD: Behind this is there a super–program or a supervisory program, which is fixed and cannot modify itself?

CS: Sometimes there is, sometimes there isn't. In the old days—1946 to 1955 or so—there wasn't. The machine was God–given. The programmer got a machine, and it had a fixed instruction set, and that was hard–wired. You could open the thing up and see the wires! And touch them, and they were hard. You could touch them and feel them and smell them.

Around 1960 things changed. Remember, there's always been a tension in computer engineering between building things that are elaborate, have many functions, and are slow and expensive, and building things that are simple, but fast and cheap. There are economies of scale, and diseconomies of scale. Around 1960, economies of scale in production and all the magical engineering stuff were pushing things towards simple, standardized, fast, cheap components. This gave us very unwieldy, inhuman machines to deal with—not that the early ones were any great bargain—but the new machines were at least an order of magnitude more powerful and effective than the old ones, and were even harder to program. And God knows, the old ones were hard enough to program. And so microprogramming was born.

The idea of microprogramming is, in fact, a meta–thing. You are given a chunk of iron—the machine. It is hard to program, but relatively cheap and fast. It is assumed that that machine will be programmed to behave like another machine. This other machine is called the target machine. The emulation is done by a microprogram provided by the manufacturer. So the guy who buys it, and the programmers who work on it, are going to see a machine which acts as though it's a certain chunk of iron, but really isn't the chunk of iron at all. It's all done with *papier maché* and colored lights and smoke and mirrors. It's really a totally different machine.

PJD: So you can talk about meta–objects?

CS: Yes. In fact they've been around commercially. The 360 series, introduced in 1964—all but one or two models, and they were strange—were microprogrammed. If you were a 360 assembly language programmer or 360 machine language programmer, their hardware was not at all the machine you saw. This meta–thing means that the machine you work with is not the machine that's actually there. From 1964 onward the machines were designed to be run with an operating system. That is, you do not read your own tapes, you do not go to the machine and press a button and start up your job, your program. You tell the operating system, "Hey, I'd like to run my program." The operating system then says, "Fine, just sit in the waiting room and read a magazine. The nurse will call you." The operating system does all the job scheduling. It does all the sequencing and scheduling of input and output. In fact, it allocates the different computer resources among the competing users, and it tries to balance them the best it can, and it does the accounting. It's like the counselor at camp—and the different programmers with their programs are the campers. It's the benevolent big brother. And it is a program. It is provided by the manufacturer. And it's *frankly* a program. It's not hidden in microcode or anything. You actually get source listings for it, and you can modify the operating system if you want to. So there's another level of metaaction that you simply have to get used to if you're going to be a programmer. If you don't like it, you'd better choose some other kind of work.

PJD: How many metalevels can you distinguish?

CS: Ah well, the levels are what you're given. There is the microcode metalevel. I cannot think of a decent-sized machine today that does not come microcoded. The hardware you get is not the hardware you're programming, even if you're at the low-

Scientific computation in 1979. (*Courtesy of NASA.*)

est possible level, so you get an operating system. That's two metalevels. Then, your own code. That's a third level. When you are a programmer, you probably don't write self–modifying programs. That turns out to be a very dangerous thing to do. It's such a powerful tool, you can do a lot of harm. It's like the old cartoon—"Amazing, it would take 4000 mathematicians 4000 years to make a mistake as bad as that." You get so much leverage that if you do even one little thing wrong, you can really destroy yourself; or, rather, destroy your program, and all its data, and every other kid on the block.

Summing it up, at the bottom level there's the machine. On top of that is your microcode. On top of that your assembly language. On top of that your high level language. Still higher are things like the special purpose packages. And all of it is surrounded by the amniotic fluid of the operating system which is supposed to keep the levels straight.

PJD: What about recursion?

CS: I wasn't brought up on recursion, and I've never been totally comfortable with it, but computer science has made recursion central these days. I think recursive algorithms are slow and ugly, and if it's possible at all to do a task non–recursively I'll choose that way. However, recursion is all the rage these days. They teach it in the computer science departments: the kids these days are being taught to think recursively. It gives me a headache, personally.

PJD: Do these metalevels result in a kind of schizophrenia?

CS: Well, no, I wouldn't go so far as to say that, but it certainly gives rise to a mode of thought that is quite different from what is usual. In fact, I think it's the distinguishing characteristic of computer people as opposed to any other kind of people. If I were at a cocktail party listening to people talk, I could pick out the computer people immediately, simply by the jokes they're making, by their particular play on words, the examples they use.

PJD: These jokes refer to paradoxes of level?

CS: Yes.

PJD: Would it be proper to say, speaking popularly, that I know something—that's one level. But if I know that I know, that's a second level, and so on?

```
C-----MENUXRF MAIN PROGRAM

C-----
C-----NOTES:
C-----
C-----   THIS MAIN PROGRAM PRODUCES A CROSS-REFERENCE REPORT FOR ALL THE
C-----FILES AND GLOBALS REFERRED TO BY ONE OR MORE BINARY MENU FILES.
C-----
C-----
      LOGICAL*1 FN(133)
C-----
      LOGICAL*1 SI(133),SJ(133),SL(133)
C-----
      LOGICAL*1 FTAB(133,200,5),MFN(133,50)
      DATA MXFTAB /200/, MXMFN /50/
C-----
      DIMENSION IGUARD(200,5),ICT(5)
C-----
      DIMENSION KKTAB(5)
      DATA KKTAB /1,3,4,2,5/
C-----
      LOGICAL*1 JDOT,JDASH, JQUOTE
      DATA JDOT /'.'/, JDASH /'-'/, JQUOTE /''''/
      LOGICAL*1 JEQUAL
      DATA JEQUAL /'='/
C-----
      INCLUDE 'IODBA.CMX'
      INCLUDE 'GINB.CMX'
      INCLUDE 'MENUB.CMX'
      INCLUDE 'DEBUGB.CMX'
      INCLUDE 'CNTRL.CMX'
C-----
C-----INITIALIZE
      ND1=5
      ND2=6
      ND3=8
      ND=3
      IDBUG=-1
      DO 5 J=1,5
      DO 5 K=1,MXFTAB
5     IGUARD(K,J)=0
C-----ASK FOR REPORT FILE NAME
10    CALL ASK ('Report file name (with extension)--|',0,2)
      IF (IOCODE .EQ. 0) GO TO 10
      CALL SCOPY (IRESP,FN)
      CALL XOPEN (ND,FN,'ASO|',132,IE)
      IF (IE .GT. 1) GO TO 80
      MC=0
C-----MENU FILE NAME LOOP
12    CALL ASK ('Menu file name (with extension) or ''DONE''--|',0,2)
      IF (IOCODE .EQ. 0) GO TO 14
      CALL SCOPY (IRESP,SI)
      CALL XUPPER (SI)
      CALL SCMP1 (SI,'DONE|',IX)
      IF (IX .EQ. 1) GO TO 300
      CALL XACCES (SI,IE)
      IF (IE .NE. 0) GO TO 16
      IF (MC .GE. MXMFN) GO TO 18
      MC=MC+1
      CALL SCOPY (SI,MFN(1,MC))
      GO TO 12
```

First page of coding for cross reference of files and global symbols in other programs. This metaprogram is written in FORTRAN. (*Courtesy of Charles M. Strauss.*)

CS: Even more than that—wondering what it is that makes you think you know. What is it that gives rise to your belief? That's the real question.

PJD: In the computer field is there some sort of principle of reduction that says that ultimately you must ignore the higher levels, and bring it down essentially to one lowest level?

CS: No. The principle, in fact, is all the other way. You want to raise it to as high a level as possible, because that's where the leverage is. Then you cascade these levels of indirectness. Mechanical things are cheap. If it has to be done by hand, it's expensive. Computer power has gone way up—computers and hi–fi equipment are the only things that are better now than they were 20 years ago.

PJD: So metathinking is in some ways a response to the high cost of labor?

CS: Well, it's certainly encouraged by the high cost of labor.

PJD: Does metathinking have any implications outside the internal workings of the computer or the computer profession? Does it have implications for the general world? Is it creating a new type of individual who works simultaneously on split levels?

CS: My answer is a qualified yes. Certainly the programmer or the computer scientist or people that hang around computers cannot help but think this way. Just as a mathematician sees quantities wherever he looks, and an engineer sees little crystals and matrices whenever he looks at metals or concrete, and physicists see atoms, and chemists see molecules, computer people see metalevels. They see data flowing and things talking about other things. A programmer is proudest when, by changing only one line of code, the generality of the code is extended enormously. That gives him a sense of enormous power. When programmers sit around in bars consuming beer after beer, this is what they talk about.

PJD: Are metalevels a part of the taxonomy of the natural world?

CS: It's certainly possible to treat much of the natural world in this way. The food chain can be thought of as a stacking up. There are all kinds of hierarchical situations. Political organizations can be thought of this way. Why New York City doesn't work anymore, or why peaceful resistance works or doesn't work, and in which cases it will work, can be "analyzed" and reduced to questions of levels of control.

PJD: In your key phrase "metathinking as a way of life," you're speaking of the computer profession.

CS: Yes, and its applicability to the outside. I was thinking of how pervasive this way of thinking is. It colors your relationships and your modeling of the outside world. it isn't only what you do at your desk or at the machine.

For example, metathinking makes me impatient with a great deal of what I read in the paper. I read that we have a Capitalist System and the Russians have a Communist System. This implies something about prices, production and the flow of goods. My training as a computer scientist leads me to suspect that I can write a computer program that is sufficiently popped–up in level so that it contains both systems within it. To run the program, I just have to set my variable C = CAP or C = COM, or set C equal to any degree of blending of the two.

Now why should any sensible individual want to engage in perpetual hostilities with the threat of blowing up the world on the basis of how one simple variable has been set? (Of course, we should make sure that the variable is set so that a plentiful supply of goods flows to me.) You get the point? Metathinking makes me impatient with the usual political verbiage that is around.

Three Meanings of Computation

THE ACTIVITY that goes by the word "computation" is now so extensive that it probably carries as many different shades of meaning as there are people who use computers. Three particular meanings will be singled out and explained:

1. The Logicians' or Computer Scientists' meaning.
2. The Physicists' meaning.
3. The arithmetic (traditional) meaning.

The Logician's Meaning. Since digital computation now embraces such diverse activities as word processing, painting a picture on a color screen and doing one's income tax, it is reasonable to seek a common basis for these activities in terms of certain exceedingly simple operations. Such a basis was provided in the mid–1930s by Alan M. Turing (1912–1954) prior to the actual construction of a real world electronic digital computer. (Several other logicians including Emil L. Post had independently arrived at equivalent versions.) Turing's abstract construct is now called a Turing machine, and a logician or a computer scientist is apt to say that the meaning of the word 'compute' is simply to call into play the operation of a Turing machine.

To give a rough idea of what kind of thing a Turing machine is, I shall follow, with suitable abridgements and alterations, the article by Martin Davis entitled "What is a Computation?" Abstraction is the process of eliminating all but the essentials. Does it make any difference whether computations are performed by a person or a well educated horse? No? Then that aspect can be eliminated. Does it make a difference whether it is done with a pencil, a pen, or with piles of pebbles? No? Well, there goes another inessential. Now look at a typical multiplication as one might carry it out on paper:

$$\begin{array}{r} 11 \\ \times 12 \\ \hline 22 \\ 11 \\ \hline 132 \end{array}$$

Does it need the above solid two–dimensional format? No, because it can be written linearly in the form $11 \times 12 = 22 + 110 = 132$. Standardize this by putting the symbols on a linear tape divided into discrete spaces or squares. Can we simplify a bit more? Yes. Think of this as a message.

1	1	×	1	2	=	2	2	+	1	1	0	=	1	3	2

Everyone knows that all messages can be expressed in Morse code, a dot–dash language. Two symbols therefore suffice to get our message through. Call the two symbols 0 and 1, and let the squares carry either of these symbols.

A tape with a sequence of zeros and ones, then, is the way the problem is presented on a Turing machine. This tape must be capable of being scanned, square by square, by moving to the right or to the left, and as the result of the scanning, a decision must be taken as to what shall be done next. A symbol can be left unaltered or it can be replaced by the other symbol. Finally, the whole process may be terminated.

In the view of Turing, then, a calculation is a finite sequence of steps, taken in a certain order, which consist of

(a) printing 0,
(b) printing 1,
(c) going one square to the left,
(d) going one square to the right,
(e) going to step *n* of the program if you read a 1,
(f) going to step *n* of the program if you read a 0,
(g) stopping.

The order in which the computer carries out these actions is governed by the specific program that is written by the programmer and placed on the tape. This may be standardized in many ways and constitutes the coding of the programming language. For example, we could choose the following code:

0 0 0 is interpreted as "print 0"
0 0 1 "print 1"
0 1 0 go left one square
0 1 1 go right one square
1 0 1 | 0 0 . . .0 1 | go to step n in program if 0 is scanned
n entries
1 1 0 | 1. . .1 0 | go to step n in program if 1 is scanned
n entries
1 0 0 stop.

A program will operate on input data. That input data may be another program or it may be the very program under which the machine is being controlled.

This, then, is how computation shapes up in the minds of the logician, and it is one of the central dogmas or axioms of the business that all computation can be broken down and then built up from such simple steps. Once a conceptual standardization has been prepared—and the Turing way is not the only way—then there is the possibility of describing its scope and of comparing its scope with that of other standardizations.

The notion of what is or what is not a computation has a considerable amount of vagueness and elasticity, and it is bound to change as time goes on. Turing's work provides an axiomatization of this concept, and ever since Euclid, axiomatization and formalization have been considered the hallmark of theoretical science. We can grant to the work of Turing (and of the other logicians of the first third of the 20th Century) intellectual status comparable to that of Euclid. Still, it would be a fair conjecture that this paradigm can as little describe what the world will come to view as computation as the axioms of Euclidean geometry can describe the real, astronomical space against which the cosmos is choreographed.

The Physicists' View of Computation. While the Turing machine provides an abstract conceptualization of computation, no physical computer operates as in Turing's outline. To get a little closer to reality, one should look at what a physicist who interests himself in the internal workings of a computer says.

The digital computer, to such a person, is simply a device that processes discrete information by performing sequences of logical operations on that information and on its own program.

By discrete information is meant strings of "status" symbols: zeros and ones or *T*s and *F*s (for true and false) or "ons" and "offs"—in short, any symbolization of two distinguishable physical states.

By a logical operation is meant those modes of statement composition such as "not," "and," "either/or," "if/then," "if and only if," by which compound statements are built up from primitive statements. Each of these modes of composition has a truth table associated with it. For example, the operation "and" has the truth table:

1st statement *P*	2nd statement *Q*	1st and 2nd *P* and *Q*
True	True	True
False	True	False
True	False	False
False	False	False

The meaning of the second line of this table is that if statement *P* is false and statement *Q* is true, then the compound statement *P* and *Q* is false.

The physicist regards the third column of this table as simply the result of inputting the first two columns (along each line), and ignoring any linguistic or logical meaning, is concerned only with how a physical device may be built which responds according to this set pattern:

"and"

Input	Output
1 1	1
0 1	0
1 0	0
0 0	0

Such a device is called a "gate," the one above being an "and" gate. In the view of the physicist, a computer is a vast assemblage of logical gates, and his job is to devise gates which operate reliably in a minimal time, to discuss their potentialities for concatenation, physical requirements, and limitations, etc. Such gates, in principal, might be mechanical, electromagnetic, electronic, hydraulic, cryogenic, chemical, or even organic, and physicists working at the production level are not concerned with the use—even to physics—to which such an assemblage of gates is to be put.

The Arithmetic Meaning. In the common or traditional usage, computation means arithmetic. To compute, to calculate, to reckon, to figure, means to add, to subtract, to multiply, or to divide. To compute, therefore, is to carry out one of the four venerable operations taught in grade school. The principal consumers of computation in this sense are business, science, technology, and mathematics itself. In the past, the mathematical demands placed on computation by business were fairly simple: do the four operations, principally addition, over and over again on vast amounts of business data. The bills sent out monthly by department stores to their credit customers are simple computations. To your balance at the first of the month add the cost of your itemized purchases. Subtract the payments made during the month and add as a carrying charge a certain percentage of the original balance. This is the new balance.

It was the business of the grade schools to teach children how to compute. At the completion of eight grades, children of my generation knew, or were supposed to know, how to add, subtract, multiply and divide (in both the short and long forms). They knew how to carry out these operations on fractions and on decimals. They knew how to reduce fractions to lowest terms, to find the greatest common divisor, and to find the least common multiple. They were also taught how to extract square roots and to apply these operations to problems in mensuration.

It will help us appreciate the revolution of computation that has engulfed the world if we remind ourselves that up until a hundred years ago, when universal education was introduced, the four basic operations of arithmetic, in their general form, were the exclusive property of a few business clerks, a few scientists, and a lawyer or a clergyman here and there. My mother, fifty years ago, could not do long multiplication or division. My father, who had four years of grade school education, had taught himself these operations out of business necessity. In our generation, I suspect, though I have seen no statistics, that if the number of functional illiterates in this country is higher than we think, the number of arithmetic illiterates is even higher.

Between my parents' generation and my own, universal arithmetic literacy became an accepted standard. Along with this came increased arithmetic complexity in our daily affairs. The paycheck of an average person working in the U.S. contains perhaps a dozen deductions, some fixed, some at a percentage, some depending upon bracket, some made periodically and some sporadically. This complexity is a vivid illustration of how arithmetic rises to saturate the newly created arithmetic computing facilities. When it comes to that famous document known as the federal income tax form, even the most tender–hearted mathematician would agree to call it a masterpiece of obscenely hypercomplex arithmetic. A lot of people literally cannot do their day–to–day sums. Most of us don't bother to check the arithmetic of the bank and payroll office. When it comes to the federal income tax, the legal, economic, and arithmetic complexities have driven millions into the offices of the tax consultants, who dedicate their best energies to keeping things complex.

Computation—the four arithmetic operations carried out with pencil and paper—is tedious, error prone, and excruciatingly slow work. Make no mistake about it: people hate to do it. That is why people have invented counting boards, strings with knots tied in them, the abacus, and clever systems for reckoning with the fingers and hands. In some places arithmetic was never carried out on paper or in the head but only with these instruments, so it may be misleading to say that difficulty with arithmetic prompted their development: no abacus, no arithmetic.

Under the pressures of computation in astronomy, in surveying, and in navigation, logarithms were invented and perfected. These reduced the more difficult operation of multiplication to the less difficult one of addition, reduced division to subtraction, and root extraction to a simple division. Varieties of analog devices such as slide rules and nomographs were devised and were found useful in their day. Astrolabes and special devices for astrological purposes were common in the Islamic world of the fifteenth century. Mechanical adding machines date from the early seventeenth century, and became commonplace by the mid–nineteenth century as standard equipment in the business world. As late as the 1940s models were still being manufactured which could only add. By the late 1930s mechanical calculators were available that performed the four operations by merely pushing buttons, and in the early 1950s commercial models were brought out with a square root button. This last was a device which could have been implemented in the early 1800s. By this time the mechanical computer was giving way to the electronic computer, which was eleven orders of magnitude faster and far, far more versatile. Think of it: eleven orders of magnitude! What other aspects of the world have been multiplied in such a short time by a factor of 100,000,000,000?

The four arithmetic operations, done in the head or on the back of a brown paper bag, are error prone, and systems of checks were devised for "proving" the results. To prove a division: multiply back. To prove a subtraction, add. But how did one prove a multiplication or an addition? There were such things around as the rule of nine which gave partial checks. In the final analysis, you proved an addition or a multiplication by doing it twice. The old fashioned grocer, took the pencil from behind his ear, wet its tip on his tongue, and wrote down the items on the grocery bag. First he checked to see whether he had as many items listed as the number on the counter in front of him. Then he totaled up the sum. As a check, he did it again. If he were really sharp, he added once going up the column and once again going down.

Possibility of an incorrect computation has not yet been extirpated by the electric computer. Have the items of a grocery sale been entered properly? If they have, then with very high probability, the sum will be executed properly. If they have not, the computer will not know about it; if you complain, then the standard answer will be that the computer made a boo–boo, but every one knows that it was not the computer but a human being that boo–booed somewhere down the line. If the price of the items is extracted from memory automatically by the reading head triggered by the zebra stripes, then how do we know that the prices have been stored in memory accurately? We can formulate an endless sequence of queries; has this been recorded properly, has that been programmed properly? Ultimately our faith in the accuracy of the whole process is based upon our faith in the accuracy of the human agencies at work, their ability to discover and correct errors by checking the reasonableness of the answers at various interim stages.

The Higher Arithmetic: Elementary Scientific Computation

If the mathematical demands of the business world could once have been satisfied by the four basic operations of addition, subtraction, multiplication, and division,—this is no longer the case, by the way— the demands of emerging science were more substantial. In planar and in solid mensuration, geometry led to more complicated operations. The area A of a square of side s is given by $A = s^2$, the volume V of a cube of side s is given by $V = s^3$. Inversely, sides in terms of areas or volumes are given as $s = \sqrt{A}$ or $s = \sqrt[3]{V}$. In this way, and already in ancient mathematics, square roots and cube roots came into the arithmetic world. The length of the diagonal of a rectangle, it has been known since before Pythagoras, also requires the computation of a square root. The computations of the parts of triangles, an important question in surveying, astronomy and navigation, requires a knowledge of the so–called trigonometric functions, and these were computed, tabulated, and worked with systematically by Klaudios Ptolemaios (Ptolemy) (150 A.D.), and perhaps earlier.

Geometry and trigonometry, then, are full of laws or formulas, expressible in simple algebraic terms, or even in terms of the common language, and each of these laws requires a numerical calculation for the evaluation of particular cases. Similarly, physics is full of laws, often simple in their formulation but of deep significance, and each of these laws implies the possibility of carrying out a related computation. One can cite many examples, but consider the law of the body falling under gravity: $s = \frac{1}{2} gt^2$, Kepler's Third Law of Planetary Motion: $\frac{R^2}{T^3} = c$, the law of relativistic mass: $m = \frac{m_0}{\sqrt{1 - \frac{v^2}{c^2}}}$, the law of refraction: $n_1 \sin_1 = n_2 \sin_2$. Planck's Law of Radiation $R = \frac{c_1}{\lambda^5(e^{c_2/\lambda T} - 1)}$. In times gone by, these formulas were evaluated with pencil and paper and with an occasional look in a book of tables, or on slide rules, or on mechanical calculators. The degree of complexity of the computation is small: a few additions, a few multiplications, and one is done. Today, such computations are done on hand–held calculators that have function buttons; indeed, the instruction manuals for those instruments often advise purchasers to learn the operations by practising with such formulas of physics.

These formulas or laws and many others like them have one unifying feature: they can all be carried out with operations of arithmetic or of the higher arithmetic. In the category of higher arithmetic I include the four elementary operations, the operations of extracting roots, the employment of the elementary transcendental functions such as the exponential, the trigonometric functions sine, cosine, tangent, and the inverses of these functions. By iterating these operations, one can build up such functions as $\sin^{-1}\sqrt{\dfrac{1+x^5}{1+x^3}}$, $e^{-2\pi x}\cos(60x+23)$, etc. Expressions such as these are sometimes called *explicit* formulas, and it is often desirable in the progress of research to obtain an explicit formula for the quantity under discussion.

Of course, there are many useful special functions which have been studied intensively and for which a method of efficient computation is well known. Why limit the higher arithmetic to the powers, the exponential, the sines and cosines, and their inverses, all of which functions were on the mathematical scene by the early 18th Century?

Κανόνιον τῶν ἐν κύκλῳ εὐθειῶν			Table of Chords		
περιφε-ρειῶν	εὐθειῶν	ἑξηκοστῶν	arcs	chords	sixtieths
∠′	ō λα κε	ō α β ν	½°	0;31,25	0;1,2,50
α	α β ν	ō α β ν	1°	1;2,50	0;1,2,50
α∠′	α λδ ιε	ō α β ν	1½°	1;34,15	0;1,2,50
β	β ε μ	ō α β ν	2°	2;5,40	0;1,2,50
β∠′	β λζ δ	ō α β μη	2½°	2;37,4	0;1,2,48
γ	γ η κη	ō α β μη	3°	3;8,28	0;1,2,48
γ∠′	γ λθ νβ	ō α β μη	3½°	3;39,52	0;1,2,48
δ	δ ια ις	ō α β μζ	4°	4;11,16	0;1,2,47
δ∠′	δ μβ μ	ō α β μζ	4½°	4;42,40	0;1,2,47
ε	ε ιδ δ	ō α β μς	5°	5;14,4	0;1,2,46
ε∠′	ε με κζ	ō α β με	5½°	5;45,27	0;1,2,45
ς	ς ις μθ	ō α β μδ	6°	6;16,49	0;1,2,44
ς∠′	ς μη ια	ō α β μγ	6½°	6;48,11	0;1,2,43
ζ	ζ ιθ λγ	ō α β μβ	7°	7;19,33	0;1,2,42
ζ∠′	ζ ν νδ	ō α β μα	7½°	7;50,54	0;1,2,41
⋮	⋮	⋮	⋮	⋮	⋮
ροδ∠′	ριθ να μγ	ō ō β νγ	174½°	119;51,43	0;0,2,53
ροε	ριθ νγ ι	ō ō β λς	175°	119;53,10	0;0,2,36
ροε∠′	ριθ νδ κζ	ō ō β κ	175½°	119;54,27	0;0,2,20
ρος	ριθ νε λη	ō ō β γ	176°	119;55,38	0;0,2,3
ρος∠′	ριθ νς λθ	ō ō α μζ	176½°	119;56,39	0;0,1,47
ροζ	ριθ νζ λβ	ō ō α λ	177°	119,57.32	0;0,1,30
ροζ∠′	ριθ νη ιη	ō ō α ιδ	177½°	119;58,18	0;0,1,14
ροη	ριθ νη νε	ō ō ō νζ	178°	119;58,55	0;0,0,57
ροη∠′	ριθ νθ κδ	ō ō ō μα	178½°	119;59,24	0;0,0,41
ροθ	ριθ νθ μδ	ō ō ō κε	179°	119;59,44	0;0,0,25
ροθ∠′	ριθ νθ νς	ō ō ō θ	179½°	119;59,56	0;0,0,9
ρπ	ρκ ō ō	ō ō ō ō	180°	120;0,0	0;0,0,0

Ptolemy's Table of Chords. (*From A. Aaboe, "Episodes from the Early History of Mathematics," courtesy of the MAA.*)

Why not include the gamma function, the Bessel functions, the Legendre, the Airy functions, etc., etc., and in this way enlarge the vocabulary of higher arithmetic? There is really no reason not to, except that the demand for these latter functions is substantially smaller than the demand for the former, so that this limitation creates a convenient conceptual unit.

The culmination of the idea of scientific computation as higher arithmetic is enshrined in a book started in the mid–fifties and issued in 1964: *The Handbook of Mathematical Functions*, produced by the mathematical staff of the National Bureau of Standards in Washington. This book is a thousand–page compilation of the properties of the principal functions useful in physics, engineering, statistics, etc. It includes tables, formulas, and graphs in great number. As is often the case with monumental works, this book was already obsolete in its planning stages, because it represented and was derived from the pre–computer numerical technology. As the human compilers selected the functions, the individual expansions, identities, and representations for inclusion in the volume, as one selects strawberries, with loving care, computers and their software were already in full and chaotic development, but the formats of computerized information were so unstable (as they still are) that experts of the late fifties could not agree on a better way of issuing the information.

As it turned out, the book was a wild success, and probably a quarter of a million copies of it have been sold in a variety of editions. It was given the affectionate nickname "The Red Monster," after its size and binding color. Though ancient in spirit, it is still a very useful book to have around.

If all scientific computation could be performed by an appeal to explicit formulas of the sort just mentioned, and if a few such computations (say ten or a hundred) sufficed to arrive at each answer, then there would hardly have been any pressure from the scientific community to develop calculators beyond the mechanical stage of the early 1900s. Unfortunately or fortunately, depending upon one's point of view, this was not the case; in order to see why, to see how numerical computation became more complex and more subtle, we shall have to talk about the problems of numerical analysis.

To some extent, what is considered to be an *explicit* solution to a mathematical problem is a matter of convention, but now that we have laid down the conventions beforehand, a surprising thing emerges: a mathematical problem may have an answer, but it doesn't necessarily have an explicit answer.

This was realized early in the history of mathematics. While one can "extract" the square root of 2, one cannot do so in such a way that the answer is the ratio of two integers. Such ratios may approximate to $\sqrt{2}$ quite closely, sufficiently closely for all applications, but strict equality cannot be achieved. This example is from the 5th century B.C.

While the first order differential equation $y' = x^{-1}+y^{-1}$, $y(1) = 1$ has an answer, this answer is not to be found among the simple elementary functions arrived at by compounding polynomial functions, rational functions, exponential and trigonometric functions or their inverses. This example is from the 1830s.

If our problem is important enough and cannot be solved by familiar tools, then mathematics progresses by fashioning new tools, which in a later generation become perfectly familiar and standard. Power series, Fourier series, orthogonal functions, special functions expressed as integrals, all are instances of this tendency. As problems become more specialized, their special theory becomes more complicated and less generally applicable, and ultimately becomes the private turf of a handful of experts. For example, a century ago a knowledge of the theory of elliptic integrals was part of the equipment of *every educated mathematician*. Therefore, a solution in terms of elliptic integrals would have been considered "explicit." Today, this is not the case. It is a dream of the creators of such computer languages as MACSYMA and REDUCE, to "automate" all of special function theory so that this specialized knowledge will be made available to the casual user. Today, a practitioner has to consider the tradeoff between the time it would take to learn the ins and outs of an advanced special theory, and any advantage to be gained in doing so.

Teachers of higher mathematics often say to their classes: this problem has an answer, but we cannot show explicitly what it is. A course in "pure" theory would concentrate on the existential aspects of this statement: prove that there is a solution, and demonstrate a few of its features. On the other hand, if it presumed from outside evidence (physical or computational) that the problem has an answer, and if it is very important to obtain the answer, and if it is obtained by a computation, then such a course is *one of numerical analysis*.

By way of example, suppose we are confronted with the initial value problem $y' = x^{-1} + y^{-1}$, $y(1) = 1$. Existence theory tells us that there is a unique answer to this problem and it is known that this answer cannot be elementary. The approach of numerical analysis is to arrive at a numerical representation of the answer by a step by step process which approximates the differential operator $\frac{d}{dx}$ (which is defined in terms of limits or infinitesimals) by a finite difference operator.

These two tendencies, the computational and the existential, historically developed side by side. Great theoretical mathematicians were often great computers, but in the period from about 1850 to World War II, most mathematical talent was deflected away from computational matters. Computation often fell into the hands of astronomers, but the rise of the high–speed computer stimulated a great resurgence in the interest of mathematicians in computation, and this increase is continuing up to the present day. Current research in numerical analysis is devoted to the development of computational strategies that are accurate, stable, convenient, rapid, and economical.

Superarithmetic

By superarithmetic—also called large scale numerical calculation—I mean the incredibly rapid performance of incredibly large amounts of ordinary arithmetic. Take audio signal processing, for example. Suppose we want to process signals that go up to 32,000 cycles per second. Suppose that each cycle is represented by 32 data points. Then we shall be dealing with 1,024,000 data points/sec. Allowing, say, a factor of 10 with which to do our filtering, we come up to about 10,000,000 operations per second.

Suppose we want to do weather predictions over the United States. Think of the U.S. as roughly a rectangle 3000 miles long and 1500 miles wide. If microclimate is defined at "stations" every 10 miles, this means 45,000 stations on a two dimensional grid. If climate up to 50,000 feet is measured every 100 feet, then we have a total of 22,500,000 stations. This vast quantity of data must be processed at time intervals that are small enough to guarantee numerical stability. We are soon in the domain of computations which require hundreds of millions of flops (floating point operations per second). A computer that can deliver this is now known as a supercomputer, and the scientific disciplines are teeming with problems that require such capabilities. To name a few: in atmospheric studies, weather forecasting with satellite data; in

Weather patterns. (*Courtesy of the National Oceanic and Atmospheric Administration.*)

astrophysics, the evolution of stellar systems; in chemistry, the quantum dynamics of molecules; in civil engineering, the optimization of structures; in genetics, the simulation of living cells.

One of the areas that has exploited great amounts of number–crunching power is that of digital image processing. The applications range from cleaning up the images that come from planetary exploration to the production of interesting visual effects for sci–fi movies. Lucasfilm of San Rafael, California has gone into this heavily.

When a supercomputer is used to tackle problems, what, in fact, is it doing, mathematically speaking? Suppose we want to compute the aerodynamic properties of an automobile, to reduce its energy requirements. Then we must solve the following partial differential equations: (here I quote from C.W. Hirt and J.D. Ramshaw, "Prospects for Numerical Simulations of Bluff–Body Aerodynamics")

$$\text{“}\frac{\partial u}{\partial t} + \frac{\partial u^2}{\partial x} + \frac{\partial uv}{\partial y} + \frac{\partial uw}{\partial z} = -\frac{\partial p}{\partial x} + g_x + \nu \left(\frac{\partial^2 u}{\partial x^2} + \frac{\partial^2 u}{\partial y^2} + \frac{\partial^2 u}{\partial z^2}\right)$$

$$\frac{\partial v}{\partial t} + \frac{\partial uv}{\partial x} + \frac{\partial v^2}{\partial y} + \frac{\partial vw}{\partial z} = -\frac{\partial p}{\partial y} + g_y + \nu \left(\frac{\partial^2 v}{\partial x^2} + \frac{\partial^2 v}{\partial y^2} + \frac{\partial^2 v}{\partial z^2}\right) \quad (1)$$

$$\frac{\partial w}{\partial t} + \frac{\partial uw}{\partial x} + \frac{\partial vw}{\partial y} + \frac{\partial w^2}{\partial z} = -\frac{\partial p}{\partial z} + g_z + \nu \left(\frac{\partial^2 w}{\partial x^2} + \frac{\partial^2 w}{\partial y^2} + \frac{\partial^2 w}{\partial z^2}\right)$$

where (u,v,w) are velocity components in the respective coordinate directions (x,y,z), p is the fluid pressure divided by the constant fluid density, v is the kinematic viscosity, and (g_x,g_y,g_z) are body accelerations in the (x,y,z) coordinate directions. These momentum equations are to be supplemented with the incompressibility condition

$$\frac{\partial u}{\partial x} + \frac{\partial v}{\partial y} + \frac{\partial w}{\partial z} = 0.\text{”} \quad (2)$$

This system of mathematical equations was developed by the mathematical physicists of the 18th and early 19th centuries, and was available long before the invention of the automobile, but its solution cannot be expressed in terms of the mathematical vocabulary of the period of its development. It must be obtained by means of superarithmetic.

Hirt and Ramshaw go on to discuss the numerical strategy for solving these equations and the computer program which effectuates this strategy:

“The numerical solution of these equations utilizes a mesh of rectangular cells with edge lengths δx_i, δy_j, δz_k, where subscripts refer to the ith cell in the x-direction, the jth cell in the y-direction, and the kth cell in the z-direction. The entire mesh consists of IMAX cells in the x-direction, JMAX cells in the y-direction, and KMAX cells in the z-direction. A single layer of cells around the mesh perimeter is reserved for the setting of boundary conditions, so that the fluid-containing region of the mesh consists of IMAX − 2 by JMAX − 2 by KMAX − 2 cells. The proper use of these boundary cells eliminates the need for special finite-difference equations at the boundaries, i.e., equations used in the mesh interior are also used unchanged at the boundaries of the mesh.

A cycle of calculation to advance the flow configuration through a time interval δt consists of the following two major steps:

1. Finite-difference approximations of the momentum equations, equation (1), are used to obtain guesses for the new time-level velocities using the previous time-level quantities in all convective, viscous, and

> pressure gradient terms. The new velocities thus obtained will not necessarily satisfy the incompressibility condition, equation (2). Therefore,
> 2. the pressure is adjusted in each mesh cell to insure that the finite-difference approximation of equation (2) for each cell is satisfied. These pressure adjustments must be done iteratively, because a change of pressure in one cell will upset the balance in neighboring cells. The number of iteration sweeps through the mesh, necessary to obtain a desired level of convergence to equation (2) in all cells, varies with each problem. However, a typical number for problems described in this paper is 5 to 20. Typically, more iterations are required to get a problem started, because large initial flow transients require large pressure adjustments. The lower iteration number is more typical as nearly steady flow conditions are approached. Of course, the iteration number drops to unity when steady conditions are reached.
>
> From a mathematical point of view, the iteration process is used to obtain the solution of a Poisson equation for pressure, although this equation is not explicitly written out in code. From a physical point of view, the pressure iteration is necessary to account for the long-range influence of rapidly propagating acoustic pressure waves that maintain a uniform density."

In technological design, say aero– or hydrodynamics, superarithmetic is the culmination of five hundred years of steady theoretical development. It will enable new design concepts to be analyzed in a matter of weeks or months rather than in years.

The goal of the next decade is to produce multi–gigaflop supercomputers (10^9 floating point arithmetic operations per second). This will be achieved through increasing the speeds of the basic components and through highly parallel processing, new programming strategies, new numerical algorithms, and a pronounced artificial intelligence orientation.

With such machines at hand, the pace of technological innovation will accordingly increase; whether for good or ill is another question.

Further Readings. See Bibliography

M. Davis (1978); P. Eiseman; C. Hirt and J. Ramshaw; H. Goldstine (1977); R. Landauer (1967, 1984); P. Lax; N. Metropolis, J. Howlett and G.C. Rota; B. Parlett

What Scientific Computation is for

THE OBVIOUS REPLY to the question is: scientific computation, oh stupid one!, is to get numerical answers to scientific problems. Perhaps this is the case when we are students and are learning the computational trade; our teachers seem, indeed, to be satisfied when we come up with the right answer*.

Richard W. Hamming, formerly of the Bell Telephone Laboratories and one of the nation's leading numerical analysts, thinks otherwise; he says the situation is much more complex. In 1962 Hamming published a book on numerical analysis—the theory of scientific computing—which carried the motto: "The purpose of computing is insight, not numbers." This motto created a bit of a stir when the book was published. It left some readers confused. Not numbers? Then why compute? To other readers, it was like a breath of fresh air or perhaps like the smell of coffee in the morning. The motto was discussed, applauded, derided, parodied, and whenever Hamming gave an invited talk, it was certain that someone in the audience would ask him what he really meant when he wrote those words.

I shall try to extract a fair statement of Hamming's meaning from pages 3–5 of the second edition of his book, published in 1973. There we find several meanings given to the motto. The first is: "Computation should be intimately bound up with both the source of the problem and the use that is going to be made of the answer—it is not a step to be taken in isolation from reality." The remaining meanings that Hamming ascribes to his motto have to do with certain difficulties that are intrinsic to the processes of scientific calculation as carried out on a digital computer: round–off error, truncation error, feedback in the algorithmic process, its stability or instability, and, finally, the existence of competing algorithms for any particular task.

* Some educators have soft–pedaled this. In their view, the important thing is that the student hold the right methodological thoughts about the problem, not that the right answer be obtained. Try that one out at the fish market.

Computation, then, is to be put back to work on computation itself, to gain insight into our algorithmic procedures and to gain confidence in the answers that the computer delivers. The tyro may think that scientific computation is an open–and–shut matter: What's the problem? Just do the operations that the formulas tell us to do! To him, the aforementioned difficulties may come as a shock. If so, he can learn something here, but it is not my intent to elaborate the craft of numerical analysis; it is, rather, to focus on Hamming's first point, that scientific computation is not done in isolation, but for real–world reasons. What are they?

We do not eat numbers, nor do we put numbers on our backs to ward off the cold. Numbers are converted to actions. Actions, reciprocally, are converted to numbers. A caterer is asked to provide food for a testimonial dinner. On the basis of the invitation list, the caterer computes how much creamed chicken he will need. He orders his cooks to prepare that much chicken. Having served the party, he is left with bills for provisions and labor and a check from the organizers. From numbers to numbers again. In the scientific line, as well, computing leads to action. A refinery is built, a linear accelerator is built, as the order is given to drill for oil at such and such a location, all as a result of certain standardized computations.

Numbers can be converted to actions because computation is a way of achieving knowledge. It is a means for transforming data, which is already an abstracted form of knowledge, into another form more amenable as a basis for action. Computation may be the basis for the transformation of one action into another. Before computerized reservation systems were adopted by the airlines, commercial planes flew millions of passengers successfully. The computer has transformed one reservation system—and its related human acts—into another. Whether it does the job better or worse, whether it is or isn't cost–effective is not the issue here. It is our way.

Computation, then, is one of the avenues to knowledge. What are some of the other avenues? Experience (or, in the word of the laboratory, experimentation), deduction, "intuition," guesswork, metaphor and analogy, and, so the theologians tell us, revelation. Avenues of knowledge amplify one another. They also compete with one another. Claims are put forth by computer people that some of these channels of knowledge may be amplified, strengthened, or avoided altogether by computing. Though some of the claims are the extravagances of self–advertisement, in some of them there is considerable truth. There is a long list of experimental physicists, engineers, biologists, whose grant proposals have been turned down under the rubric: experimentation no longer needed. Compute!

Consider, as another example, the role of aerodynamic experimentation. Airplanes have traditionally been designed on the basis of a lot of computation based on much experimental information obtained in wind–tunnel tests. Now wind–tunnel tests are exceedingly expensive, and are themselves only partially reliable as a source of data, because the limited airstream in the tunnel does not accurately represent what takes place when a plane flies in the open air.

Since all the equations of aerodynamics are known, why not merely solve these equations by computer? Then we would have lowered costs and greatly augmented flexibility. We would have constructed the "computerized wind–tunnel"! This is still only a dream, at the time of writing partially fulfilled. What difficulties intervene? Some of the difficulties are those mentioned in Hamming's second group of explanations of his mysterious motto: round–off error, truncation error, numerical instabilities. To these add that the complexity of the computation (i.e., the total bulk of number–crunchings) is enormous, and that in some aerodynamic regimes we do not yet have adequate information on what mathematical model should represent physical reality, and what digitalized model should represent the mathematical model. Nonetheless, many authorities believe that the solution of the problem is within the grasp of the next generation of ultra–large, ultra–fast computers. If this is indeed the case, then the wind–tunnel will be an obsolete experimental device, to be placed alongside Benjamin Franklin's electrical machine in the Smithsonian.

Computation thus serves the purposes of improving itself not only, as Hamming emphasizes, with regard to the difficulties of the processes of numerical analysis, but also with regard to the difficulties of the formation of models of physical or social processes. One proposes a mathematical model of a process. What validity or utility does it have? Put it on the computer. Run the model, and see if it predicts how the world really behaves. If it does, good; if not, change the model and repeat.*

* In fiscal year 1983, $6,800,000 was allocated by the Department of Energy (one of numerous government agencies so doing) "for Analytical and Numerical Methods. This research addresses questions basic to improved predictions of the feasibility, reliability, safety, and efficiency of energy systems. Equally important, this research

Surely there must be limits to how much knowledge can be coaxed out of computation, unless one believes—the mystic principle of "the macrocosm in the microcosm"—that the whole world exists, in miniature, inside the computer.

Once a successful computational scheme has been worked through, the operation of the computer may then proceed automatically; we may pass without human intervention from action to numbers and back to action, as when a computer is hooked on–line to a piece of factory equipment or to a prosthetic device. Realizing this, we might well imagine that the whole machine has come into being without the conscious intervention of mathematics, for it certainly exists in the real world, and who is to say that what has proved possible is possible only through one avenue of development? The mathematics is the product of the human imagination, and the computer becomes a concrete realization of part of it. It is to be likened and contrasted with the flowers, crystals, planets, atoms of the natural world which, in their behavior, are also mathematical machines *when we choose to impute mathematics to them.*

Thus far we have gone on the assumption that our mathematical model is in place, or nearly so. This model, if it comes from the physical sciences, is itself the fruit of hundreds of years (some would say thousands) of the best scientific thought. If it is true that space flight would have been impossible without the use of computers, it is equally true that the machines could not have been set a–crunching without the classical equations of motion. The work of the great scientists: Ptolemy, Brahe, Kepler, Galileo, Newton, is a precondition for computation.

Is it possible that computation can bypass this kind of thinking, and by working somehow with the raw materials of nature come up with its own formulation of the relevant laws, and ultimately derive from these laws whatever information is held to be of utility? "Heresy!" would be the cry of one group of scientists. "It is possible" is the answer of another.

addresses questions basic to the fundamental laws of physics describing the structure and properties of matter, focusing on mathematical and computational techniques. Analytical and Numerical Methods comprises three categories of research activities reflecting the principal thrust of the program in previous years: Applied Analysis, Computational Mathematics, and Numerical Methods for Partial Differential Equations." (*Summaries of the FY 1983 Applied Mathematical Sciences Research Program*, DOE/ER-0157/1, page 3).

Here is the judgement of physicist Stephen Hawkins, Lucasian Professor of Mathematics in the University of Cambridge, in his famous inaugural lecture of 29 August 1980:

> "At present computers are a useful aid in research, but they have to be directed by human minds. However, if one extrapolates their recent rapid rate of development, it would seem quite possible that they will take over altogether in theoretical physics. So maybe the end is in sight for theoretical physicists if not for theoretical physics."

On which Clifford Truesdell, a profound scientist and historian of science, remarks that if physicists, "their intellects already wan and flagging from the ravages of malignant computeritis," go this route, they will be committing hara–kiri by computer.

By way of summary, scientific computation serves a variety of purposes. To the nonscientific laity, its internal workings must be ignored as it leads silently from action to action, but the actions themselves should be placed under the most severe scrutiny. To scientific researchers, it is a tool, one of many, for eliciting knowledge, to be used as the occasion necessitates. To the philosopher, it raises many issues, including the ancient one, whether Pythagoras was right when he asserted that all is number, and the new one, whether, paradoxically, by virtue of digital computers, mankind may soon ignore the numbers.

Further Readings. See Bibliography

R. Hamming; C. Truesdell

Why Should I Believe a Computer: Computation as Process and Product

TO AVOID high–flown generalities here, I shall discuss a particular problem. This problem will be broken down into steps or phases and in this way the implications of the title of this section will become clearer.

The problem is one in acoustic engineering. Let us suppose, along with E.N. Gilbert of the Bell Telephone Laboratories (see Pierce, p. 149), that we are interested in computing the reverberation time of a concert hall. This quantity plays an important role in the acoustic quality of the hall.

1. The Real World Situation. This consists of the concert hall, with its size, its shape, its materials, the disposition of its furniture, the audience, the kind of music that is being produced, and whether it is to be amplified or not. To some extent, this information may be described in a natural language such as English. Here and there we would want to supplement ordinary English by the more precise spatial language provided by geometry.

2. The Mathematical Model. The laws of physical acoustics are invoked and a mathematical equation is derived which is thought to describe the physical situation. Gilbert has fixed on a certain integral equation. The process of arriving at a model inevitably involves a simplification of the real world situation. It may assume a knowledge of certain physical constants which are difficult to obtain theoretically or experimentally. The language of the mathematical model is that of classical mathematical physics.

3. Strategy of Solution. The mathematical equation is sufficiently complicated so that a solution cannot be written down in elementary terms. A digital computer is required, and a method of approximate solution must be devised. The space and time variables, which in the mathematical model are continuous, must be discretized; that is, space and time are both divided into many small intervals. Among various competing strategies for solution — approximate solution, really — one particular strategy must be selected. Gilbert has selected a certain iterative process.

The language of the numerical strategy is that of a portion of classical mathematics known as numerical analysis.

4. The Computer Program. Once the numerical strategy has been determined, it is possible to write a computer program to carry out the numerical analysis.

The language of this phase is some particular computer language good for numerical analysis and conveniently run on the available computers. This might be FORTRAN, ALGOL, BASIC, APL, etc.

5. The Computer Output. The button is pushed, the computer grinds away and ultimately puts out answers. The language of the answers may be in the form of numbers, graphs, or other symbols.

It is now up to the acoustical engineer to react and take steps on the basis of this output.

The whole process of steps one through five may be iterated several times.

Steps one through five provide a fairly general outline of a certain class of engineering computation. A few more steps might have been separated out, but these will do for our discussion. What are some of the difficulties that may intervene between inception and the computer output? Here I leave Gilbert's specific work and talk generally.

At step one, the scientist may have described the physical set–up erroneously or may have failed to take into consideration some important feature.

At step two, the mathematical model selected may be an inadequate expression of the physical process.

At step three, the discretization process may lead to a numerical algorithm exhibiting unsatisfactory roundoff or instability properties,

Partial Scheme for a
Technological Computation

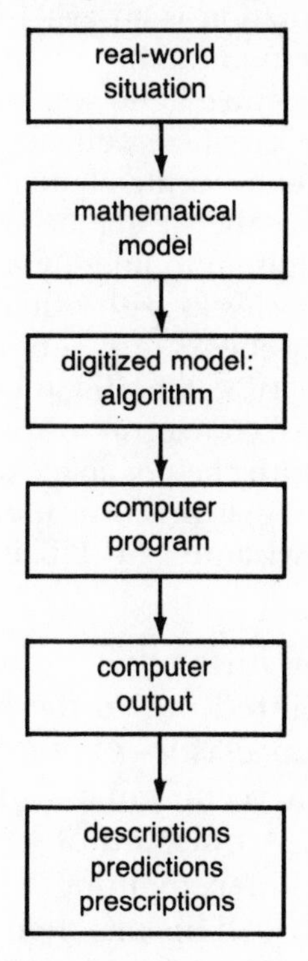

leading to numbers which in no way can be considered close to the true theoretical solution of the mathematical model.

At step four, the computer program may be an erroneous translation of the numerical algorithm.

At step five, for a variety of reasons, the computer may not do what you think you have programmed it to do.

At every step or sub–step there are opportunities for errors or inadequacies of widely different types to enter and to vitiate the answers obtained.

To elaborate some of the difficulties associated with steps three and four, I will quote the words of Professor Ian Sloan of the University of New South Wales, one of the leading numerical analysts in the world:

> "It is not sufficiently appreciated that it is extraordinarily difficult to come up with the right answer at the end of a long calculation. My

> Though scientific computation is an art, I believe that a proper approach to scientific computation can be taught. One aspect is, of course, the importance of well–structured and well–documented code. (Sometimes it is even worthwhile sacrificing efficiency in the interests of creating a simpler and more easily understandable code.) But even a well structured program can give the wrong answer, so a proper attitude to checking is essential. In addition to line–by–line numerical checking, I place great stress on cross checks with other work, special cases, and finding internal consistency checks (e.g. sum rules). In fact, I approach checking as a creative activity, calling for much imagination in the devising of checks. And then, of course, there are the problems of rounding errors, for which theory (with the possible exception of linear equation solvers) seems not yet of much practical use. In practice one is often forced to rely on the empirical test of stability under appropriate perturbations."

To elaborate some of the difficulties associated with the whole process, I cite an example pointed out to me by my colleague Professor Herbert Kolsky. A uniform elastic–plastic beam is pinned at its two ends and is subjected to a rectangular pulse of pressure uniformly distributed over its length. A question of some importance is whether the final plastic deformation remaining in the beam will be in the direction of the applied pulse. This problem was run with ten different programs, such as ABAQUS, ADINA, ANSYS, etc., much used in structural analysis. Six of the programs predicted a permanent deflection in the direction of the load and four in the opposite. One of the programs required 2.9 hours on a Cray 1S (a supercomputer). The correct answer is not known at the time of writing. (P.S. Symonds and T.X. Yu, "Counter Intuitive Behavior in the Problem of Elastic–Plastic Beam Deformation," *Journal of Applied Mechanics*, 1985.)

This example can be multiplied many times over. Tell it to a numerical analyst and he will laugh; this kind of thing is what keeps his profession in business, what prevents his experience from being made redundant by packages for scientific computation. The recitation of these difficulties and of isolated "horror" stories, however, should not blind us to the past record of solid achievements in many areas.

In view, then, of many possibilities for error, why should I believe a computer? What are my criteria for the acceptance of the computer output?

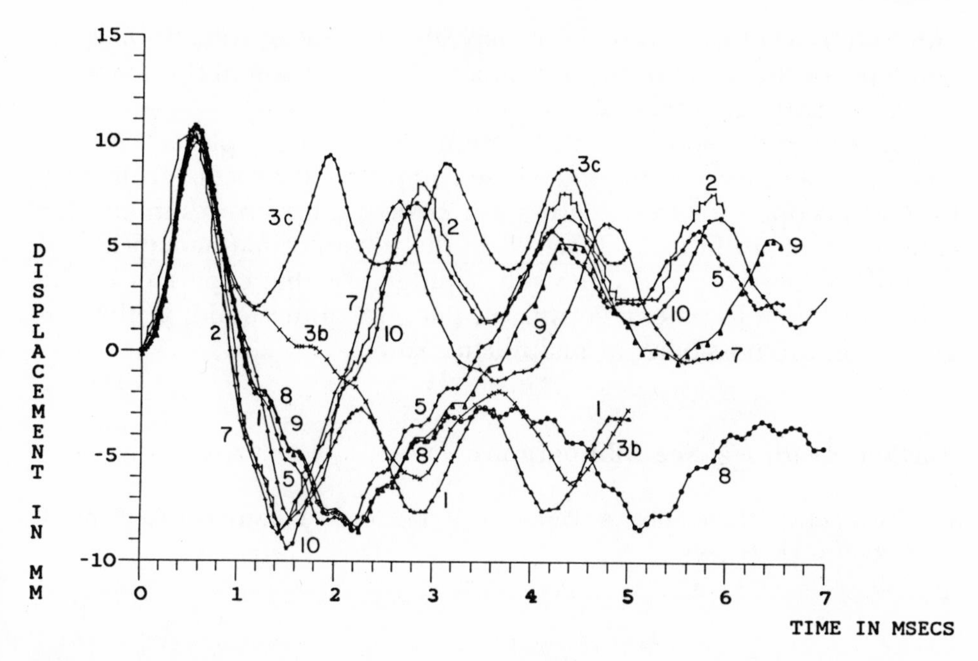

Different programs provide different answers. Which is correct? (*Courtesy P.S. Symonds.*)

The answer that I can give, and which I think is the only sensible answer, is this: in the first place, many of the individual steps and sub-steps have been tested out in a variety of situations that, while not identical to the one under scrutiny, have many features in common with it. Some of the purely mathematical parts have been amplified by mathematical proofs that their behavior is as claimed. Secondly, the whole process from start to finish can be regarded as delivering a product. This product is designed to do a certain job, and the criteria for the acceptance of, or the successful completion of, the job lies outside the product itself.

One criterion may be that the computation duplicates what has been determined experimentally. (It must be pointed out that experimental results do not always deliver what the scientist who has designed the experiment thinks they will deliver. A discussion of the sources of experimental error would take us too far afield here.) The product may be in competition with other products designed to do much the same job but whose computational steps may differ radically from the one under scrutiny. Thus there may also be the possibility of independent confirmation of the output and a comparison of the economy of the competing processes.

I have used the language of the commercial market place, but what I really mean to say is that the whole computation process, from start to finish, is open for examination, for criticism, for improvement, for experimental verification. It stands in the forum of mathematical and scientific experience and ultimately is judged by that experience. The process is open to patent, copyright, sale, warranties, and, in this litigious age, the possibility of malpractice suits.

Further Readings. See Bibliography

P.J. Davis (1972); R. de Millo, A. Perlis and R. Lipton; E. Dijkstra (1976, 1978); D. Johnson; J. Pierce

The Whorfian Hypothesis: Ends and Means in Computer Languages

Computer Babel

"And the whole world was of one language and of one speech." Thus is it written in Genesis 11:1. The interpreters of the Bible say that the generation following the Flood was fearful lest the world be destroyed again, so they set out to build a tower whose top would be so high that people could survive there above the crest of any future flood waters. This project was not pleasing in the sight of the Lord. He confounded the language of the workmen, and the construction was frustrated. This explains the diversity of natural languages. It is an early instance of the Nostalgia of an Imagined Unity.

I sit in front of my word processor and insert my thoughts into it by means of a program known commercially as EZRITE. Underlying this program, whether in hardware or in firmware, there are an assembly language, a machine language, and an operating system language. If we limit our attention to business and scientific applications, and exclude the sorts of underlying languages just mentioned, there are today more than 800 computer languages and dialects in which fair amounts of programming take place. Perhaps 95% of all computing takes place in the business language COBOL, reflecting the fact that the great bulk of computation being done is business data processing. For scientific computation FORTRAN, BASIC, APL and Pascal are all popular languages.

It is evident, then, that now there is no, nor was there ever, unity among computer languages. The machines constructed in the 1940s were each "spoken to" differently, according to their particular construction. Still, as with natural languages, the dream of unity and all the benefits that unity might confer, persists. Though most observers believe it to be an impossible dream, it does energize useful changes.

Perhaps the most unified language that the world now knows, with a sphere of influence that cuts across all the natural languages, is the language of mathematics. Once the natural language barrier is overcome, the same symbols, the same interpretation, allow a mathematician in Russia to communicate mathematical ideas to a mathematician in Papua, New Guinea. One might think, then, that computer languages, which derive so much from the mathematical spirit, would reflect this unity.

The difference is this: mathematics is written in a mixture of formal and informal (natural) languages. Despite what some people think or claim, mathematics has never been totally formalized; In contrast, computer languages are set up with an extraordinarily high degree of formalization, which must deal with all manner of trivialities and tiny details. For example, the integral is one of the great, enduring concepts

a vector $E_p \neq 0$ in $\mathfrak{g}^{\beta+p\alpha}$, and put $E_n = (\text{ad } E_\alpha)^{n-p} E_p$ for $n \geqslant p$. Then $E_n = 0$ for $n > q$; if $p \leqslant n \leqslant q$, then $E_n \neq 0$ as a consequence of (iv) and Theorem 4.2 (v). We shall now prove

$$[E_{-\alpha}, [E_\alpha, E_n]] = \frac{(q-n)(1-p+n)}{2} \alpha(H_\alpha) E_n \qquad (n \geqslant p). \qquad (7)$$

Since X_β is a scalar multiple of E_0, (7) would imply (ii). We prove (7) by induction and consider first the case $n = p$. By the Jacobi identity

$$[E_{-\alpha}, [E_\alpha, E_p]] = -[E_\alpha, [E_p, E_{-\alpha}]] - [E_p, [E_{-\alpha}, E_\alpha]]$$

$$= 0 + [E_p, H_\alpha] = -(\beta + p\alpha)(H_\alpha) E_p,$$

which by (i) equals $\frac{1}{2}(q-p)\alpha(H_\alpha) E_p$. Now assume (7); we have

$$[E_{-\alpha}, [E_\alpha, E_{n+1}]] = -[E_\alpha, [E_{n+1}, E_{-\alpha}]] - [E_{n+1}, [E_{-\alpha}, E_\alpha]]. \qquad (8)$$

The first term on the right is $[E_\alpha, [E_{-\alpha}, [E_\alpha, E_n]]$ which by induction hypothesis equals

$$\frac{(q-n)(1-p+n)}{2} \alpha(H_\alpha) E_{n+1}.$$

The last term on the right-hand side of (8) is $-(\beta + (n+1)\alpha)(H_\alpha) E_{n+1}$ since $E_{n+1} \in \mathfrak{g}^{\beta+(n+1)\alpha}$. Using (i) we find that these two terms add up to

$$\tfrac{1}{2}\alpha(H_\alpha) E_{n+1}\{(q-n)(1-p+n) + p + q - 2n - 2\}$$

$$= \frac{(q-n-1)(n+2-p)}{2} \alpha(H_\alpha) E_{n+1}.$$

The language of mathematics. (*From S. Helgason, "Differential Geometry, Lie Groups, and Symmetric Spaces," courtesy of Academic Press, Inc.*)

in mathematics. It is a concept at the macro level, and the integral sign and its accompanying notation emerged historically and persisted as a unifying symbol of this concept. In contrast, the concepts that underly precise formalization, whether in mathematics or in computer languages, are petty in nature, residing at the micro level, and are much more subject to the personal whims of the individual formalizers.

Practical necessity in designing computer systems requires one to make comparisons between different computer languages. Within the comparative study of natural languages there is the "Whorfian Hypothesis," which says that the structure of our language determines our perception of the world. (Benjamin Whorf, "Collected Writings," John Carroll, ed.) Whorf's position is opposed by Noam Chomsky, who asserts that we are all born with a universal language facility which is due to preprogramming of the brain.

The Whorfian Hypothesis: Ends and Means in Computer Languages

PJD: What is the Whorfian hypothesis?

CS: I wouldn't call it a hypothesis; it's an idea, a great idea. Benjamin Whorf seems to have popularized it. It says that some languages are better for some things than other languages are. It can be summed up in the phrase: "If you don't have a word for it in your language, you'll find it difficult to talk about it." I found examples of this in *The Greeks*, by H.F. Kitto. In the early chapters Kitto talks about Greek being a language with a spurious clarity. As an example: I gather from my one year of Greek that it's natural to say in Greek, "on the one hand . . . on the other hand." This is so natural to Greek that it's hard to organize a sentence without it—natural thesis and antithesis, or natural dilemma. Now, a language that doesn't have this particular construction so deeply ingrained in it is not going to enforce the balanced point of view, the dilemma. It might enforce a single point of view; it might encourage a multiple point of view; but Greek is particularly good at this "on the one hand . . . on the other hand."

PJD: So Greek, artificially almost, dichotomizes the world.

CS: Yes. French is, in some ways, a very clear language. German is a very unclear language. It's very easy to be vague and metaphysical in German simply because a natural way to express yourself in German is to write one sentence and then another, and run them all together with all the verbs piled up at the end.

Every natural human language has its own bent, has its own things that it's good at. I've heard it said that in the Eskimo tongue there are a large number of words for snow, because it can be very important to an Eskimo if the snow is light and fluffy or coarse and granular. In Arabic there are 27 words for camel: pregnant camel, old male camel, that sort of thing.

The same holds for computer languages. I'm talking here both about the basic machine language, that which the machines are built to understand, and the so–called higher level languages that have grown up in the computing community because the basic machine language is very difficult and awkward for a human being to use. All of these languages, however artificial, are languages in the sense that you can issue commands in them and get things done. They all have their natural bent, and what I mean by the Whorfian hypothesis applied to computing is that, depending on the language you choose to program in, you will find some things a whole lot easier to do than other things. Some things are going to be easy and some are going to be hard *no matter what language you choose.*

PJD: How much option does the average programmer have in the selection of the language?

CS: It depends on where he works. If he works for a standard business/data processing-oriented place, not that much. He has Cobol; he might have PL/1; he might have FORTRAN, Pascal, assembly language. These days Ada is being tried out. Each of these was, in its time, a new, magic language designed to make everything wonderful. Well, just as there ain't no magic human language, there probably ain't no magic computer language.

PJD: There are dreams of a sweetheart language?

CS: There are always dreams of a sweetheart language, just as there are always dreams of sweethearts! Hope springs eternal. It turns out that we don't know that much about language.

PJD: What about trade–offs when you're building or designing a language? Do you end up with having to sacrifice this in order to get that?

CS: Invariably. In the late fifties and early sixties people were very excited about what should go into a language. Everyone was inventing his own language, because that was the first wave of higher level languages, compilers, and compiler compilers and all that kind of stuff. Then people got tired of it. It was like the Hundred Years War. All people wanted to do after it was to be left alone. They got tired of fighting and decided that yes, your God might be as good for you as mine is for me.

Recently, because of the rise of Ada and Pascal and distributed processing, parallel processing, and a whole lot of commercially available things in computing, not wild–eyed researchers' dreams but real, funded things with millions of dollars worth of project and equipment money behind them, the debate about what should be in a computer language seems to be showing distressing signs of rising again. You don't want it too rich, or you will spend all your time learning it, and you don't want it too poor, otherwise you'll spend too much time trying to say the computer equivalent of "I need a cup of coffee."

PJD: Does a new language arise because programmers, over a period of time, find themselves doing the same operation over and over again, having to program it, and say, "If I just had some magic word to give me this, it would be great?"

CS: No, that's too simple. I can't think of a single computer language, except maybe BASIC, which doesn't have subroutine facilities. So any repetitive process is one of the first things you think of in a language—how to take a repetitive process and in effect make it a command. Set it off, and you never have to look at it again. All you have to do is get the inputs right, and the right outputs will come out. That's standard.

The new features arise first from new hardware features.* You must realize that all of the computing industry has been pulled along by the engineering. Neither the computer scientists,

* See the comment at the end of the section for an alternative view.

business, and Defense Department nor any of the other usual movers of technological development have had much effect on computing. They like to think they have, but they really don't. If you look at what the pull is from, where the real developments are from, they are from software being sucked along in the way, in the vacuum, behind computer hardware development.

PJD: Is there any indication that computer languages are settling down? Over the last 40 years they have been very unstable, changing much faster than natural languages.

CS: In fact they did settle down for a while in the seventies. Pascal, which came in in the seventies, was a reasonable success, but the seventies was basically a time of consolidation. Then towards the end of the decade, the new architectures came along: networks, distributed processing, and parallel processing, and the federal government also realized that the old ways were no longer going to do.

There's an interesting combination of business/psychological/inertial problem with what the government did. The DOD has a lot of obsolescent equipment and computers, and computer programs hanging around, and they realize that they're going to have to junk all of that very soon and get new stuff, because no one's going to be around to service the equipment any more, and they won't be able to get parts for it. They also realize that reprogramming all of this stuff is going to be an incredible job. What they want to do is have a language that's going to make the reprogramming of the old stuff and the programming of the new stuff a lot easier and more maintainable. It's easy enough to write code; what's tough is maintaining it. That's what Ada was developed for.

The jury is still out about Ada. Ada is very nice in a lot of ways, but it still might be a little too complicated for the average guy to program. It's like a natural language with 17 case endings! It was designed by a committee of computer scientists. You've probably heard that a camel is a horse designed by a committee.

PJD: How much money has been spent on Ada?

CS: A very, very big bundle.

PJD: How do we realize what we want to say? You said that the Whorf hypothesis is that if we don't have a word for it, we can't express it. How do we know, at any stage, what it is we want to say?

CS: Well, that's one of the big troubles, because you don't want to say it unless you think of saying it, and in programming you usually don't think of saying it until someone asks you to say it. So there's a vicious circle there—they aren't going to ask you to perform impossible tasks. So in lockstep, creeping forward, the state of the art is extended. Then people's expectations rise to meet the state of the art, and they find practical uses for it.

There are a lot of short steps, and a lot of communications problems, because the Joes on the firing line, namely the programmers and the analysts, find it hard to get the word up to the guys who design and implement languages. It's like machine tools or progress in the construction business or any place where there are guys in the trenches who do the work, and they are not in direct contact with the guys who give them the tools with which to do the work. That's what conferences are for, that's what papers, that's what research is all about. That's something that computer science is supposed to be good at. I don't think it *is* very good at it.

PJD: In natural languages one occasionally sees a certain amount of political or social manipulation of the language. For example, in Marxist writing there is a certain specialized vocabulary, and everybody recognizes it. In the women's movement there is revision of the English language to get rid of sexist expressions.

CS: That's a very good example of the Whorfian hypothesis, or what the feminists feel to be the Whorfian hypothesis.

PJD: Does one ever see what one might call a political maneuver within computer languages?

CS: Everyone has his bag of tricks, like bidding in bridge. If you have a standard problem to face, a sort, or a scan, a breaking up of a line of text into words separated by blanks, things like that, everyone has his own favorite way of doing it. It's habit, and, in a particular situation, any one of five ways is as good as another. Rather then spend ten minutes thinking about which one is marginally better, you just do it the standard way. It's not only natural, it's actually a timesaver—the microseconds you'd save by doing it the very best way are not worth the time you'd spend worrying about which was the very best way.

The basic techniques you use—not so much how you manipulate things—but how you store the data, whether it's stored in trees, or stacks, or tables, or random access files or sequential access files, and whether you store it formatted or unformatted—all that kind of stuff is a matter, I think, of training and experience. I feel particularly strongly about this because I was never really taught storing things in trees. I've read about it, and I can do it, but it's not natural to me. And I still work according to the little dictum that "If we're able, we'll use a table. If you insist, we'll use a list." Now that's a very old–fashioned attitude, and kids in introductory programming courses these days are taught to use stacks, and algorithms that use stacks. Pascal is a very natural stack language. It's built of stacks and trees and things like that. I was a member of an earlier generation. I can use them, but it isn't natural. You show these kids algorithms using tables and they say "Heh? Why do it that way; what a stupid way to do it! Why not use a tree?" A lot of it is personal preference, but it's more than just that, it's the result of a whole programming style. To use them, of course, you've got to have languages that have the constructs. It is very difficult to use trees as your data structure in FORTRAN. There's no word for trees.

PJD: Let's focus a bit on this aspect of language. Language comes in as an answer to a perceived need. Then people develop it, and the language turns around and changes those very needs. For example, I imagine that this would have occurred in the field of accounting. Accounting procedures were already well established in the 1940s, at the beginning of the machine era. Then machines were used for accounting, languages were built, and so on. Did those languages then turn the accounting procedures around?

CS: No, you picked the one area that has been impervious to the computer revolution. The 1980 revisions to the tax code were a very good example. I had to recode some depreciation schedules, and to do that I had to find out what the rules were. So I looked it up in the code, and after 20 pages of incredibly inscrutable text I got to the meat of the thing, where they say in words, long involved words, how it's done, and then they give me a table of examples. I tried to work through the examples, and I didn't get the numbers in the table. Why is this? I called up my friends at the IRS, and they called up their friends, and they called up their friends back down the line. It turned out they rounded off to make it easy in the table. The table lied.

Now, this is not my idea of the computer age. What they should have is a little algorithm, because everyone's going to have to program this. The reason we programmed it for this particular project was that all our competitors had programmed it too. Everyone had to go through the same translation from the old–fashioned language of accounting to the new–fashioned language of computing, and it suffered in the translation. You know the phrase "it loses in the translation"—this definitely lost in the translation.

PJD: Ultimately with a new generation of accountants coming along who are familiar—

CS: This is already a new generation of accountants.

PJD: And they still have a conservative practice?

CS: Very, very, very, very conservative.

PJD: How do you suppose a computer scientist who's in the business field would view the income tax forms that the average person has to make out?

CS: In fact, there aren't any computer scientists in the business field, or at least not academic computer scientists. There are two kinds of computer languages: those that are good for mathematical manipulations (FORTRAN, PL/1, APL)—although they have their proponents for business data processing—and business data processing languages. These are not terribly good at complex mathematical operations. They're intended more for data manipulation, data retrieval. They suffer greatly from not having a good notation. The reason for this discrepancy is that the mathematical manipulative languages have a thousand years of western mathematical notation to draw on. Some of the greatest minds of western civilization for the last thousand years (that is, if you really like mathematicians) have been working at it, refining the notation, deciding what it is they want to say and what would be a neat way to say it. If you want to computerize that, all you have to do is copy it. You could have no better model. But in previous times no one ever thought about business data processing. It's all internalized by the profession and so there is no good notation.

I remember enough from my days as an applied mathematician to know that if you have a good notation, you need nothing more. You need no thought—that's what the calculus is all about. Before they invented the calculus you had to be Galileo or Newton to do the physics problems. Now any sophomore in college can do them.

PJD: Well, we've talked about areas of resistance. Do you know of areas where there has been great feedback into the problem area itself from the language?

CS: The development of generalized data bases and their access languages. That certainly has been an area where the users and the "used"—sorry, the users and the providers—are locked together symbiotically, and there's a lot of communication back and forth. That's primarily in business data processing.

A very active and interesting area is networking—hooking lots of little computers together randomly. It's very nice to be able to buy computers incrementally. You have another task, so you buy another computer to deal primarily with that task, but of course it has to communicate with all the other little computers you have around doing all those other little tasks. Or suppose you buy a piece of hardware and write some software for it, and it's perking along merrily. It's working so well, and everyone around it has gotten accustomed to having things done by machine, so you want to extend it. You don't want to replace it, because it's fine for what it's doing. You just want to add capacity. Now, IBM did very well with this by letting you upgrade their central processors quite easily. (All you had to do was pay them much more money.) But every time they did it, new ground rules came in, and that was bad. What you want to do is hide the ground rules from the people who are already there. You don't want to upset things that are being done successfully. It makes much more sense to be able just to buy a new chunk of computing power, plug it in, and have it do both the new task it's supposed to do, and communicate with all the other guys. That's what networking is intended to accomplish, and there has been a lot of work on how to do this. Now this isn't, strictly speaking, computer languages, although a network protocol *is* a language, at the very least a proto–language.

PJD: So there are areas in which there is a balance between what you want to do and the languages which permit you to do it.

CS: Well, at least there's feedback. It isn't as if the language is handed down, God–given, from on high, and you like it or lump it. Usually some group invents a language. There are lots of problems in inventing new languages. How do you know that the language isn't ambiguous? If it's a natural language it doesn't matter if it's ambiguous because human beings can tolerate a huge amount of ambiguity. In fact, that's what they're really good at! That's why we aren't going to be replaced very soon by computers, because the computer is going to grind to a halt if there's the least little bit of ambiguity in a language. Or even worse, *not* grind to a halt, but oscillate wildly, or do other very strange things. Human beings would say, "wait a minute!"

So in the invention of a new computer language—at the very least you have to be sure it's unambiguous. Now, that's kind of a negative virtue. Making sure that something has no possible ambiguity doesn't get the job done. It simply ensures that there are no potential errors in the job. Then someone in the group is worrying about the human factors in the language. That is, what is to be expressed and how easy it is to express in that language? Then there are boring details, like how easy is it going to be for a translator for this language to be written. That means, how easy it is to translate from the statements expressed in the new language into the basic machine language that any particular machine is able to understand? Now, some languages are going to be easier to translate than others because, although machine languages are rather similar, they do differ from each other, and machine architectures differ slightly one from another.

Well, there are a lot or problems there, and a lot of questions. And who's to say that this committee has the Jovian knowledge to do it all well, especially when you realize that they probably are thinkers, not doers. Where are they going to get the input from the "doing" community: that these here are the real problems, and those there may look like problems, but they aren't really, we know how to get around them.

In effect it's field work. Someone has to go out and look over the shoulders of the working stiffs who actually face the tasks and somehow get them done or don't get them done and abstract back from all of this local stuff. When you're doing, you hardly have time to think, too. It's a very unusual person who can *do* and at the same time stand back and *think* about what he's doing. And it's an even more unusual person, who can do, stand back and think about what *he's* doing, see what his fellow men are doing, and then stand back and abstract *that*. It's really a very difficult task.

PJD: A process such as de–ambiguization of a language—does this occur at the planning stage of the language?

CS: All throughout.

PJD: Does it occur as a debugging phase?

CS: From the moment of inception they try to keep things unambiguous. At every stage of formalization of the language—writing out the specifications of the language—there are checks. Not, unfortunately, computer checks, although there is thriving research into how, given a language, you can automatically make sure that it's unambiguous. The proof is really when the language undergoes use by large numbers of users. Just as the truth of a mathematical theorem is really tested when you publish the result in a paper, and let the world take a look at it and see if they can poke holes in it, see if they can either duplicate it or break it.

PJD: What's a simple instance of a language ambiguity?

CS: You want an actual one or a made–up one?

PJD: Either way, but a typical one.

CS: It's hard to think of a typical one. Rules for when variables can be local or global. You have a name for something. Who knows that name for that thing? What other statements in the program can know the name for that thing?

PJD: Do these ambiguities relate to differences in meta–level?

CS: They can. Typically they do, because otherwise it would be rather easy to pick them up. And again, what we don't have is a good meta–formalism, or formalism for talking about meta–things.

PJD: Is meta–thinking particularly difficult?

CS: Yes, everyone I know finds it particularly difficult. We're good at ordinary thinking. You look at a piece of code, and say, "I see it all before me." When you add another level and say, "I am now looking at a bunch of code whose job it is to look at code," right away you exponentiate the difficulty.

PJD: Talking about growth, I observed that in the early days of MACSYMA, which is a symbolic algebra–manipulation language, that the user guide book was very small. Then, as the language grew in depth and richness, as you could do more and more things, the guide books grew accordingly, and now the dilemma is whether it is easier for an individual to know all the identities of algebra and formal calculus, or to know the contents of the guide book whose function it is to get you through MACSYMA. Now, surely there's some moral there.

CS: Well, the moral is that difficult problems are difficult. And having a computer as assistant is all very well, but then "to whom much is given, from him much will be expected."

Comment: There is an alternative view of the forces that lead to the development of new computer languages.

"There are certainly languages and language features which have arisen as a result of hardware development. For example, FORTRAN was a response to hardware floating point arithmetic. The current work on parallel languages is being driven by development in VLSI (very large–scale integrated circuits). There are also hardware innovations which came about as a response to software. Built–in machine instructions for manipulating stacks were an outgrowth of languages which supported recursion. The LISP machine is an outgrowth of LISP. But most language developments are, I think, entirely independent of hardware, or nearly so.

"Where do they come from? I think that here we have to distinguish two kinds of development. There are new language features which are added, and there are new languages which represent basically new approaches to the problem of computing. New language features are added for exactly the reason suggested above: the programmer gets tired of programming the same thing over and over. But 'the same thing' is not simply a subroutine; it is at a higher level of abstraction. To take a simple example, you cannot add 'while' loops to FORTRAN or APL just by writing a 'while' subroutine.

"Really new language approaches emerge when the programming community finds that something it has to do is conceptually simple but cannot be simply described in the presently available programming languages. For instance, one of the major insights of LISP was that it would be very convenient to have a language which explicitly manipulated symbols rather than numbers. A major innovation of Ada is that it permits one to use types explicitly. And so on. These are the real creative leaps in programming languages."—E.S. Davis

Further Readings. See Bibliography

L. Miller

The Programming Milieu

PJD: What is there about the profession of computer programmer that makes it unique?

CS: Basically, it's people who would much rather communicate with machines than people. It's supposed to be a highly cerebral occupation, and I guess it's also supposed to be both forbidding and dull.

PJD: What does a day in the life of a programmer look like?

CS: It's systems. It's both pure thought and dull hack work, and they're inextricably bound up. You have to have a maniacal attention to detail when you're actually writing the code. When you're planning the code, you have to have a far–ranging set of ideas that keeps all the parts of the problem in your head. First you abstract the essentials of the program and then you start fleshing it out. As you're developing it, you may not have to keep it all in your head at the same time. You do jot down notes, and you build up a document trail, and you have to have command of that entire document trail. It's like working on a construction project where you are the architect, the engineer, the construction boss, and the construction worker all at the same time.

PJD: How does the programmer keep track of the network of details? He doesn't rely on a flow chart, does he?

CS: You never rely on one thing in particular. Flow charts are helpful, as is drawing block diagrams on what are called system flow charts. The actual code is not all that important. What's much more important is what information you need to use and how it's going to get from one place to another. If you show me code, however well documented, I'm not going to be able to make head or tail of it. If you show me a description of the information that goes in, what comes out, and any temporary information created and stored during the translation process from input to output, then I will understand the program pretty well without seeing a line of code.

PJD: Then a good programmer goes through a descriptive phase before starting to write code?

CS: You go through it, and then you have to keep it in your mind continually as a working document. You've got to know all of it and have it instantly recallable, *as you are programming.* And then you have to be able to clear almost all of it out of your mind, so that you can concentrate on the little, itsy bitsy section of code that you're generating. It's like being a painter—you have to be able to keep in your mind, "the big picture," and, yet, when you're working on some detail in the lower right hand corner, you have to really concentrate on that detail but not overdo it in one way or another. It can't be too sloppy, and it can't be too detailed.

PJD: How does the day shape up—can you split it into some kind of percentages?

CS: Mostly, when you see programmers, they aren't doing anything. One of the attractive things about programmers is that you cannot tell whether or not they are working simply by looking at them. Very often they're sitting there seemingly drinking coffee and gossiping, or just staring into space. What the programmer is trying to do then is get a handle on all the individual and unrelated ideas that are scampering around in his head. The computer itself is made up of very standardized, very simple components, and the function follows the form here. Really, any profession has only a certain number of tricks in its bag. For example, doctors have a bunch of cures. When you go to a doctor, hopefully you go in with some disease that will match one of those cures. Similarly, programmers have certain tricks in their bag. They know how to use a lot of techniques, and when a specific problem comes in the door, they can put together (and that's the key phrase) in some way enough of these facilities and tricks to solve the problem.

Putting things together is the whole art. All the rest can be taught—it's technique. It's like building something out of Tinker Toys or Legos or Erector Sets. If you happen to have gotten a big Lego set for Christmas, you have a lot of pieces, and they're all standardized. The "play" is how you put the components together. Now, the programming components are not stacked as tightly as a wall built with bricks or Legos. It's much more like Tinker Toys, where you have round pieces with holes to hold spokes, and spokes of varying lengths, and that's about it. When you see anything made out of Tinker Toys, it looks very spindly; it looks, in fact, like a skeletal outline, a network. That is exactly what a computer program is. There are chunks of code held together by transfers. You tell the computer to start executing over here in this chunk of code, and then, upon the arising or the detection of a certain condition, the computer should go over there to another part of the program and execute another section of code, and then return willy–nilly to where it transferred from in the first place; or, upon detection of some other condition, go to some other place. That is a network. The job of the computer programmer is to make sure it all hangs together.

PJD: What happens when a bunch of programmers are assigned to work on individual parts of a larger code?

CS: In that case you not only have to make sure that *your* part works, you have to make sure that, when plugged into the larger network, the interfaces (good computer buzz word) are completely well–defined and work exactly as you and everyone else intend them to. The responsibility for what goes on in your part of the code is yours. I find that part the most fun, and I think most programmers find it the most fun, because it's all yours. You get complete control of the structure of your part of the code. There are rules of good coding practice and rules of sloppy practice, but within these general rules, you have a high degree of freedom. However, at the interfaces where your section of the code begins, and at the output end, where execution of your section of code is terminated, there's supposed to be some data sitting there, and it had better be in the right form, or the next guy's section of code is not going to be able to handle it correctly. Now, the specifications for this interfacing have to be written by someone, usually by a group of programmers, and they have to be communicated without any ambiguity, and the whole plan perceived without any ambiguity, by the entire team of programmers. That's where the fun starts, because programmers really would rather deal with machines than with people. No matter how formalized a description you have of what the interface between one section of code and another should be, there's always a certain amount of human communication involved. That's where the "you did," "I didn't" sort of schoolyard bickering goes on—at a very childish level—simply because of the different connotations of words.

PJD: I've heard people speak of the *n* squared (n^2) buildup of trouble in a group of *n* people.

CS: That's because there are *n* squared (n^2) possible connections between *n* people. Judicious design will isolate the building blocks as much as possible to minimize the amount of interface between these blocks. There is only a certain amount of human capacity for understanding any section of a problem at one time. It's analogous to the direct span of control. The army squad, patrol, or battalion structure is based on how many people one person can really command directly, and I think it's 4 or 5, at the outside.

PJD: Does that hold for the programming milieu also?

CS: Yes, very much so. Which means that you can do very well as an individual programmer on an individual project up to a point, and then you get overloaded and your effectiveness goes way down. You have to get another guy on the project, but that's not going to double your effectiveness. In fact, you're only going to increase it by about 20%, because some of your time and some of his time is going to have to be spent in communicating with each other. You put in five people, and not only are they going to have to communicate a lot with each other, but you're going to have to appoint a boss who does nothing but keep the people in touch with each other. If you start increasing the scale of the project so that there are levels and levels of supermanagers and coordinators, there will be very few people actually doing the work. Granted it is a pyramid, but it's a pyramid with practically vertical sides. Which is too bad, but that's the way it is.

PJD: Is there a certain paranoia that exists among programmers?

CS: What they do is ephemeral. You can't touch it, smell it, or taste it. When it's done, it's just bits in a machine. The bits are very powerful, in fact unreasonably powerful, because without the code the machine is an inert lump, while with it the machine is a very powerful tool. That is part of the joy of programming—that you can feel this immense power. It's like driving a Maserati, or a bulldozer, or riding on the back of an elephant. However, while you're developing the program, the program is out there. It's either sitting on your desk—pencil marks on paper—or it's in the machine, and you're testing it. Text editing, file and retrieval handling systems are all built into modern–day operating systems. We see whole offices, entire business operations, converting all of their data and entrusting it to the care of this electronic device. However, you will find that programmers are among the last to adopt this new technology, although they're the purveyors. They're its cutting edge. What programmers want is listings; they want their program printed out on paper, and they want that paper locked in their desks. Now, they're never going to use that paper. The working documents are always on line files. There are times when there are practical reasons for having your programs printed out. It's much easier to make marginal notations and draw circles, and, again, since a program can be thought of as a network or, rather, a bunch of holes tied together, having it laid out on a large sheet of paper so that you can physically make the connections by drawing circles and arrows and that sort of thing can be useful. However, the use of these pieces of paper, when you can flash the whole thing up on a CRT much more expeditiously and easily, is anachronistic. It's anachronistic in rational terms; it's not at all anachronistic in psychological terms. As your program is developing, you have pieces of it that are away from you. If they get too far away from you, you've lost them. You have to spend all of your time trying to get them back, and sometimes you may never get them back. Conceptually, not physically. Physically they're there, but they're just a lot of strings. They aren't tied together yet. And, in fact, it is the programmer's job to tie them together.

I've found, and I guess a lot of other programmers have found, that the best way to do it is to relate the psychological sense, the connections in the network, to a physical representation. You know what this thing is on a certain page, and you know what the page looks like—you can conjure up a picture of it in your mind's eye—and this other part that it's supposed to connect to, transfer to, or interact with is on this other page—and you have a picture of that in your mind— and since it's printed, it's not going to change its spatial relationships. You know where it is. The trouble with information retrieval systems is that you can call things up in any order.

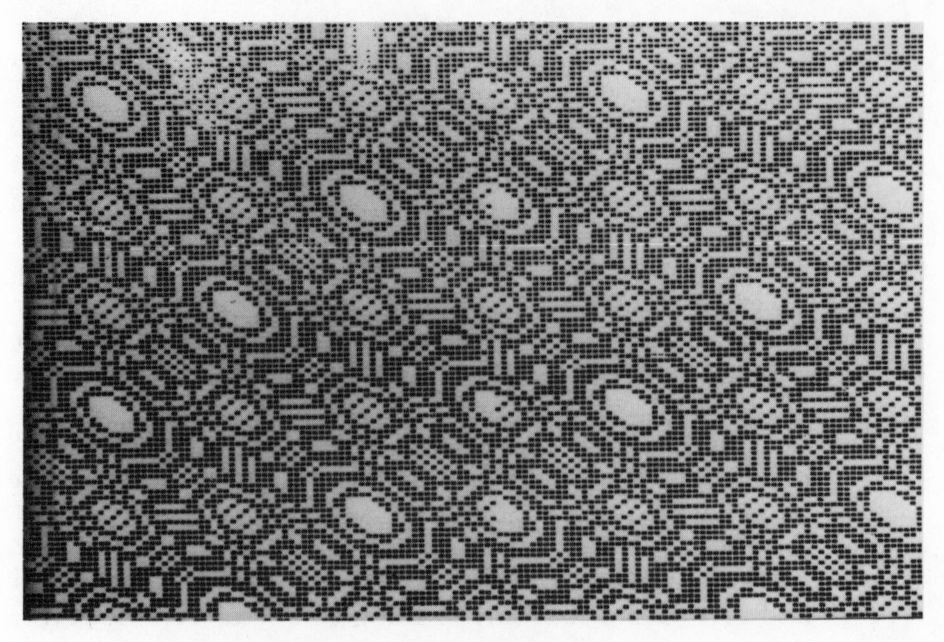

Quadratic residues. (*Courtesy of Charles M. Strauss.*)

The advantage of the paper is you *cannot* call them up in any order. You can, if necessary, fold them so that they're juxtaposed, whereas before they were separated by fifty sheets of paper. So the order can be changed. Still, there is a linear order that is impressed simply because you're printing the stuff out on paper which is linear.

PJD: These are craft skills that have to be developed?

CS: Yes, it's a craft skill, like being able to put down a smooth layer of mortar just the right consistency for a bricklayer. This is where the rubber meets the road. And I have not seen this skill in any other discipline.

PJD: You have spoken of the preference of a programmer to communicate with a machine rather than with people. Are there other psychological qualifications for good programmers?

CS: Not everyone can become a programmer. Not everyone wants to. If you don't like doing crossword puzzles, I don't think you would like being a programmer. The same sort of mental muscles are involved.

PJD: Is a mathematical talent useful in a programmer?

CS: It's useful, but it's not terribly necessary. Usually good mathematicians are bored by programming. They feel that the mathematics they know is so powerful that they don't need the machine's assistance.

PJD: What about physical skills? For example, if a person is a good woodworker is this likely to be helpful?

CS: No. In fact, I think programmers by and large have at best only an average range of physical skills.

PJD: Has anybody set up a diagnostic test to identify the ideal programmer?

CS: Yes, but all it measures is a fairly routine set of mental abilities. There's a certain thrill that a true programmer feels when working on a program. It's the exhilaration of working with, and of coupling yourself to a device (like a car, for example) that will enable you to transcend your own physical limitations. You get in a car, and you are part of it, and it is part of you. You, together with the car, can do things that the car couldn't do alone and you couldn't do alone. And they *pay* you for it, too. That's the marvelous thing about it.

PJD: There's a literature on the psychology of the Grand Prix driver. Is there a comparison here?

CS: I think there is. It's just that people haven't got onto the fact that there is a comparison.

PJD: The driver and his car are one.

CS: The programmer and his computer are not really one, because one computer serves so many programmers. However, the programmer doesn't think of it that way. The programmer thinks of it as his machine, no matter how many other people are using it. As far as he's concerned, it's his alone, and the only time he ever notices all those other people is when they get in ahead of him and he has to wait. Fred Brooks points this out very well in an early chapter of *The Mythical Man Month* when he says that the attraction between the programmer and his computer is the closest thing to poetry that you can imagine. You make this code up. Before you programmed it, it didn't exist. Even when you have programmed it, it still doesn't really exist. It's just symbols, but if you put those symbols into the computer they come to life. You get synergy. That's a true thrill. It's what keeps programmers going.

Further Readings. See Bibliography

F.P. Brooks, Jr.; S. Turkle

IV

PERSPECTIVES THROUGH TIME

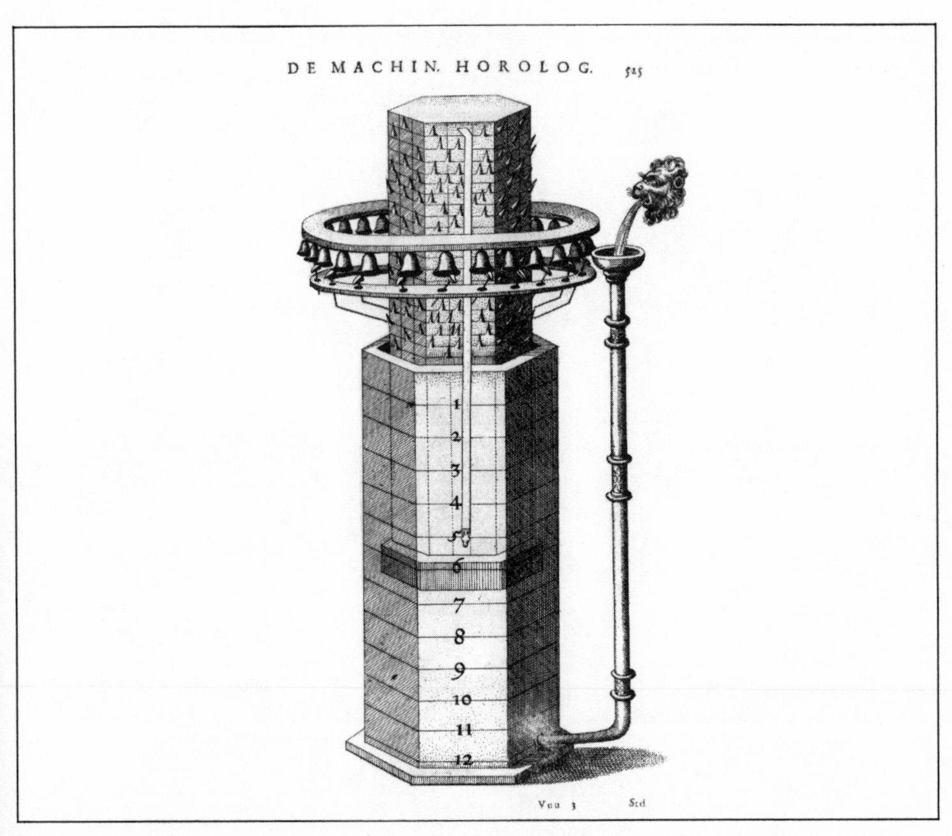

17th Century scheme for an alarm clock. (*Courtesy of the Lownes Collection, John Hay Library, Brown University.*)

Of Time and Mathematics

"He (Giambattista Vico) became convinced that the notion of timeless truths, perfect and incorrigible, clothed in universally intelligible symbols which anyone, at any time, in any circumstances, might be fortunate enough to perceive in an instant flash of illumination was (with the sole exception of the truths of divine revelation) a chimera. Against this dogma of rationalism, he held that the validity of all true knowledge, even that of mathematics or logic, can be shown to be such only by understanding how it comes about, i.e., its genetic or historical development."—Sir Isaiah Berlin, *Vico and Herder*.

"What happens when a new work of art is created is something that happens simultaneously to all works of art which preceded it." —T. S. Eliot. Quoted in Leonard B. Myer, *Music, the Arts, and Ideas*.

Of Time and Mathematics

Time, that mysterious something, that flow, that relation, that mediator, that arena for event, envelops us and confounds us all. What is time? The answer of St. Augustine has become famous: "If no one asks me the question, I know; but if one should require me to tell, I cannot."[1] Two millennia later, two revolutions in physics later, we can still sympathize with this answer. Our shelves are filled with formulas and speculations, and we still cannot say what time is; we cannot agree whether there is one time or many times, cannot even agree whether time is an essential ingredient of the universe or whether it is the grand illusion of the human intellect.

There are thus two conflicting opinions about time, and they have been around since antiquity. According to Archimedes (and to Parmenides earlier still, for whom ultimate reality is timeless), one must eliminate time, hide it, spirit it away, transform it, reduce it to something else, to geometry, perhaps. Time is an embarrassment. According to Aristotle (and to Heraclitus earlier still, for whom the world is a world of happenings), one must face time squarely, for the world is temporal in its very nature and its comings–into–being are real.

Modern science has largely followed the path of Archimedes rather than that of Aristotle. Time is downplayed, ignored, transformed, eliminated. Cause and effect are replaced by description and relation: do not ask why, but how; the successes of the Archimedean program characterize our scientific civilization.

The degradation of time in theoretical science seems paradoxical. After all, a major task of science is to describe what is and to predict what will be. Is not time locked into this aspiration? To survive, we must anticipate, and by anticipating we gain some control over raw nature. And to fulfill our incessant demand for prediction, do not our equations display t, the symbol for time, over and over again? Yet Emile Meyerson, a philosopher of science, years ago pointed out that as science has become more rational, it has tended to suppress variations in time.[2] It looks for the "constants" of nature, for the things that are invariant and hence timeless: the speed of light, the rest mass of a proton, the gravitational constant. It sets up conservation laws of mass, energy, and momenta, which tell us that when all the bookkeeping has been done, nothing has happened. Time is eliminated or, if not eliminated entirely, stands degraded.

Variables throw rules into flux. Norbert Wiener used to remind his classes of the croquet game in *Alice in Wonderland*, in which the balls were live hedgehogs and went their private ways, the wickets were playing cards and more than usually unstable, and the mallets were live flamingoes with wills of their own—and, to top it off, the rules were constantly changing. Without invariance, remarked Wiener, there can be no science.[3] To succeed in science, simplify, ignore; ignore time, for instance.

When Albert Einstein was confronted with Meyerson's observations, he admitted their truth, allowing that characteristic of thought in physics is the tendency in principle to make do with "space–like" concepts alone.[4] Aspiring to the high deductive and aesthetic perfection of Euclid's *Elements*, theoretical physics geometrizes time by conceiving it as a straight line. One speaks of "points" in time, or of "intervals" of time, and these expressions are stronger than metaphor. We work and reason with a time axis along which we display such things as variation of temperature or prices. It is so much second nature that we would be at a loss for alternative representations. We use this model even though it contravenes what seems to be a basic feature of time: its irreversibility. With our eyes we can go back and forth over a time axis from 1790 to the present, but, short of getting into H. G. Wells's Time Machine,

we cannot pull such a stunt in human time. In relativity theory, in the subtle fusion of time and space known as Minkowskian space–time, the space dimensions seem to lord it over the time dimensions, and the whole structure exists as a frozen manifold outside of time. The emphasis of general relativity is on what is space–like rather than on what is time–like, and this fulfills, in part, the dream of the English mathematician W.K. Clifford of describing matter and motion in terms only of extension.

"Is the future already here?" asked Phillip Frank, philosopher of science. In moments of fanciful speculation, when we try to create alternative systems of time, we think of time as a circle, or of time as a Möbius strip, or of time as another manifold of topology in which Frank's query might not be a paradox—and the famous time warps of television fiction put the seal of popular approval on this geometrization.

As I write, I am reminded of several contemporary textbooks on mathematical physics. Time, they usually explain, by way of introducing the differential equations which follow, is a primitive notion which cannot (and hence need not) be defined. Time is represented by a real

Endless Ribbon From a Ring I, by Max Bill. (*Hirshhorn Museum and Sculpture Garden, Smithsonian Institution.*)

number t. A particular value of time, t, is the date or epoch. An interval of time is an interval of the real number system when ordered by the relation $\leq$. To the authors of such books, time is thus a real number. It is not extracted from a mathematical structure which is discrete or granular. It is not a vector quantity or a complex number, nor is it drawn from a non–standard system that admits the infinitely small. It has the topology of the real line, though there is vagueness as to whether one deals with $[0,1]$, $[0,\infty]$, or $[-\infty,\infty]$. This depends upon which cosmological church one adheres to. At any rate, when the mathematician now places time within the abstract real number system, we have a further degradation. The geometric element is introduced by this identification and, in turn, is replaced in favor of the symbolic, structural, formalistic element, and this process of space degradation has been going on in mathematics since Descartes coordinated geometry.

Until ten or fifteen years ago, then, one could truthfully have said that the science of physics was timeless. The paradox is that most physical laws prescribed how some physical system would evolve in time. These laws were timeless in the sense that the predicted evolution would be the same whether it commenced today, yesterday, or 100 million years hence. The laws refer to time, but their description of time evolution is itself timeless, or, if you prefer, changeless.

The ongoing revolution in cosmology, however, popularly referred to as the "big bang theory," has brought about a radical change in this state of affairs. Today physicists say that the universe originated about 18 billion years ago, in an initial state when the density of matter was infinite throughout the universe. They describe the evolution of the universe, from a quark–antiquark gas in the first 10^{-8} seconds, to an electron–positron gas in the next 10^{-4} seconds, and then a radiation–dominant plasma for about 300,000 years. This was succeeded by today's mass–dominated regime, in which we observe stars and galaxies, and even planets, of which one at least (so far) is capable of supporting life.

These statements are not timeless; rather, they are dated, whether in microseconds or in megayears. Cosmology and related areas such as elementary–particle physics thus become historical sciences, or, if you prefer, evolutionary sciences, like geology and biology. The laws of physics now become in a sense temporal, for these laws are applicable only while the universe is in a state wherein the phenomena being described by the given laws can actually take place. The observable phenomena of physics in a quark–antiquark gas are utterly different from those on the surface of the earth as we know it today. In this sense the physical laws that govern today's earth—meteorology, say, or geophysics—are not timeless or eternal. Rather, they are valid only during that portion of the universe's history in which planets and atmospheres are physically possible.

The common philosophy of mathematics says that personal, historical time is absolutely irrelevant for mathematics. Some authors have even expressed the opinion that mathematics is the *one* subject in which time is irrelevant. The entities of mathematics are envisioned as timeless, existing perfected in a world of pure essences. The truths of mathematics are truths forever, outside of time, outside of mind and personality; the deductive dialogues take place atemporally in a world of pure logical transformation. Technical words in common mathematical discourse which seem to betray a temporal basis, words such as "is," "exist," "let," "vary," "approach," "map," "construct," and "equip," are held to be metaphorical expressions of a formalized time–free equivalent.[5]

Mathematics makes use of implication: $(A \rightarrow B)$ & $(B \rightarrow C) \rightarrow (A \rightarrow C)$. If we had to verify this in a specific case, it would take a moment to

Primary cesium standard NBS-6. The sixth generation of atomic frequency standards at NBS (National Bureau of Standards), NBS-6 is 6 meters long and provides a frequency reference with an accuracy better than one part in 10^{13}. When operated as a clock, this device keeps time with an accuracy of about 3 millionths of a second per year. It is part of the NBS Atomic Time System, which provides the national standard unit of time interval, the second. (*Courtesy of the U.S. Department of Commerce, National Bureau of Standards.*)

think about it and to agree. If a computer were set to verify the symbolism, it would take only a flash, but, still, a flash is not instantaneous. Yet the common philosophy has it that implication involves no passage of time.

The notion of a function is a creation of post–Renaissance mathematics. In Galileo's *Two New Sciences*, one sees it emerging from the idea of motion. In the minds of such authors as Newton, a curve is the path of a point moving in time. The notation $y = f(x)$, with its black–box–like character, stresses the individual y as the output (or correlate) of the input x; and having inputted one x, we go on to input another. The contemporary functional notation, which became popular about thirty years ago, $f{:}X \rightarrow Y$, stresses that the whole domain X of values x is mapped onto the range Y atemporally; the relationship $X \rightarrow Y$ is that of static completion.

When one writes X for a set, say the set of real numbers, this set is thought of as being "all there," simultaneously or atemporally, for us to "act upon" atemporally. And where are the elements of X said to reside—atemporally—that we may act upon them? The answer is various: in God's Empyrean, in Plato's kingdom of essences; or are they in Popper's World III, which is the world of the products of the human mind? J.C.C. Smart has described this position in no uncertain terms: "We can say, '2 + 2 is equal to 4' not because we wish to be noncommittal about the temporal position of 2 + 2 as being 4 but because it *has no temporal position at all*."[6]

The computer, insofar as it solves the equations of mathematical physics, insofar as it is an instrument of oracularity and purports to tell us today what will happen tomorrow, collapses time by making the future appear now. Not entirely, of course, because computation is a process requiring a finite lapse of time.

I have described degradation of time in the physical sciences, in the mathematics of physics, and within the operation of the formalism of mathematics itself. I turn now to a consideration of change in mathematics and to our perception and understanding of it; these are things in which the action of time cannot be denied.

One can very obviously make historical statements about mathematics. Archimedes, in the second century B.C., proved that $3\frac{10}{71} < \pi < 3\frac{1}{7}$. Lobatchewsky produced an alternate geometry in the 1830s, and Cantor introduced set theory in the 1880s. The theory of functions of a complex variable was not known to Al–Khowarizmi. The truth of the Riemann Hypothesis was not known to the world's mathematical community in January, 1985. It is clear that the creation, discussion, utilization, interpretation, and promotion of mathematics all take place in time. As time progresses, new statements are added, old errors are corrected, new concepts developed, contexts enlarged, new applications found, new interconnections and interpretations made, and old inadequacies overcome. This leads to the notion of mathematics as an agglutinative subject; Herman Hankel's view is that, in most sciences, one generation tears down what another has built on. In mathematics alone, each generation builds a new story to the old structure. Alas, I do not believe this is an adequate description of our subject. Surely, building goes on, but there is tearing down also by error correction, by simplification and abstraction, by reorganization, by the neglect of what has gone out of fashion or of what in other ways is judged irrelevant, and by the inability of the individual practitioner and of the whole mathematical community to deal with the entire historical corpus of mathematical material.

In Hankel's day, it would have been quite possible for an individual to know essentially the whole of mathematics as it then existed. Though extensive, the knowledge was within the capabilities of an individual. The last generation for which this statement would have been true was comprised, in my view, of those trained in the period 1900–1910. An individual mathematician can now hardly know, in any deep sense, more than five percent of the mathematical corpus. If it is legitimate to talk about what the whole mathematical community keeps in the forefront of its mathematical consciousness, then there are vast areas of information, potentially available, that are ignored, abandoned, condemned as trivial or irrelevant, or discarded for reasons of fashion or aesthetics. If it is legitimate also to talk about the "mathematical frame of mind" in which the practitioners worked, as individuals or in a group, then this frame of mind would include not only the specific knowledge but also the intentions, motivations, intuitions, interconnections, metaphysical principles, and values. These frames of mind are lost and are irrecoverable. When a man dies, says the Talmud, a world is destroyed.

So there is mathematics, and there is the history of mathematics and the mathematical experience. They may be linked through the question: are the truths of mathematics independent of time? Well, there must be a sense in which they are. I will not assert that while two plus three was five in the days of old Pythagoras, now, due to certain adjustments that have to be made, it is slightly more. After all, mathematics should not be a game played with live hedgehogs and flamingoes. But construing "truth" in a larger sense, the truths of mathematics are time–dependent. The mathematical author of *Alice* perhaps wrote more wisely than we suspect. There are many ways of exhibiting this dependence upon time. Indeed, let us consider for a bit this incredibly simple statement: two plus three equals five. In ancient Babylon (c. 550 B.C.) this statement would have been written in a line of cuneiform (omitted here) and would have been pronounced "2 ana–3 ūsibma 5." If we translate these grunts into our kinds of grunts, they come out, "two to three I added: five," but is translation from one environment to another ever possible without loss? Poets say it is not.

Today, I would write this statement in arithmetic notation as $2 + 3 = 5$. The shapes of the number symbols date from the fourteenth century or somewhat earlier. The plus sign dates from 1481, while the equality sign is from 1557. At an elementary level, the meaning of this sentence has to do with the counting of aggregates. If one juxtaposes two items with three items, one will then count five items; here is a statement about physical acts in the world. On the other hand, if I write two trillion plus three trillion equals five trillion, then this is not a statement about counting, for these quantities are beyond the potentialities of being counted physically. Therefore, there is such a thing as counting without counting, and the equation $2 + 3 = 5$ must contain within it (or it has imposed upon it) a non–counting context. Meaning has been extended from one environment A to an overlapping environment B by change of context. By making the two contexts agree in the overlapping portions of A and B, we force consistency and coherency in our subject.

One has to fall back a bit. *Verum et factum convertuntur:* What is true and what is made are convertible, writes Vico;[7] we demonstrate geometry because we make it. This aspect of mathematics should be stressed because it is usually underemphasized. Still, it is only part of the picture, for we are not free to make our mathematics in a totally arbitrary manner.

In time, the number system was gradually enlarged, and today a student in a course in real variable theory may interpret the statement $2 + 3 = 5$ as referring to part of the real or complex number systems. These systems were hacked out laboriously over a period of thousands of years. Is this particular line of context–development complete or even completable? Probably not. A recent addition, the non–standard number system, was created in the 1960s in order to clarify the old intuitions of the founders of differential calculus.

The sentence $2 + 3 = 5$ may reside elsewhere; for example, in the set of integers reduced by multiples of five. In such a case, we are inclined to write $2 + 3 = 0$, having identified 5 with 0: $5 = 0$. Or it may reside in Cantorian set theory, where we may write $2\aleph_0 + 3\aleph_0 = 5\aleph_0 = \aleph_0 = n\aleph_0$ for any positive integer n, so that $2 + 3$ equals whatever integer one likes, in this context.

In seeking an explanation of the sentence on the basis of mathematical logic, one might follow the path of Russell and Whitehead and arrive at it after hundreds of pages of the most detailed logical deduction. Or one may arrive at it after 10^{-9} seconds on a fast computer. On a computer, the two and the three and the five are part of a system of numbers which is neither Babylonian, nor Arabic, nor the set of fractions, nor the set of real numbers. The computer may vary slightly from model to model and the number system may exhibit such mathematical unconventionalities as $(a + b) + c \neq a + (b + c)$ or the equation $1 + x = 1$ may have many solutions. Does the world collapse because the computer may fail occasionally to obey the associative law of addition? It depends on the context; the world probably won't collapse, but a bridge might.

The Pythagoreans said that two is the female principle and three is the male principle and five is the first integer which combines both male and female. If we think this is nonsense, we must nevertheless rest assured that it was a serious matter to them. The Chinese, for their part, thought of numbers as symbols which expressed the qualities of spatial arrangements of objects.

If I think of two and three as dollar values attached to goods, then, in a grocery store, usually, \$2 + \$3 = \$5. But then again it may not be so, if the manager gives me a discount for buying several items. Or I can write 2 ft/sec + 3 ft/sec = 5 ft/sec if the velocities have the same direction; if they don't, this answer is false. In my view, the statement $2 + 3 = 5$ cannot be understood in isolation; its comprehensibility resides, in part, in the fact that it is one statement extracted from a coherent family of statements, such as $2 + 4 = 6$, $3 + 2 = 5$, etc. It is part of a scenario.

To contemplate what it would mean to have isolated mathematics, suppose I were to tell you that the statement $u^*/u//u^* > \int$ is of a revolutionary mathematical character but that it relates to absolutely nothing in the literature. What meaning can this have? It is clear to me, then, that the meaning or meanings to be attributed to the statement $2 + 3 = 5$ depend upon the circumstances under which the statement was made.

Here one might object, saying that counting is part of the primitive intuition of addition, and intuition of mathematics is not mathematics. It might be said that the embedding of the integers in the real and complex numbers is part of the history of mathematics, and the history of mathematics is not mathematics. It could be claimed, too, that bags of groceries or vector velocities are part of applied mathematics, and that is not mathematics. One might argue that the digital computer is part of physics and numerical analysis, and that is not mathematics. One may then assert that there is still a formal meaning to $2 + 3 = 5$ which can be abstracted from all of the above, and this formal meaning is out of the range of time. In that case, I will invite anyone to tell me what that timeless meaning is and to tell me in a way and in a language or metalanguage which itself is beyond the range of time. In my view, this cannot be done, because the meaning of $2 + 3 = 5$ must be supplied as part of a wider, similar set of utterances, and this meaning is bound up with application, with intuition, with arrangement, with computation, with art, with mysticism—in short, with the whole mathematical experience. The more of these elements that are stripped away to arrive at a pure, clear statement, the harder it becomes to communicate what remains and the closer we are to a formalism in which thought becomes separated from meaning.

Mathematics embraces all these aspects, and even more. "Pure mathematics don't exist," asserts Didier Norden,[8] and I agree.

Let us move, then, to a mathematical example in which the concept of time dissolves an absurdity. After the laws of signed arithmetic quantities had been worked out—positive times positive equals positive, and negative times negative equals positive—it became crystal clear that there is no number equal to the square root of -1. To put it another way, the square root of -1 does not exist. If one asks a run–of–the–mill hand–held calculator to compute the square root of -1, the lights will blink as a signal that something is wrong. Yet, as every high–school student of algebra knows, there is a square root of -1. After the work of Cardano (1501–1576), its formal aspects were well known. By the seventeenth century its utility within mathematics was well known. By the nineteenth century not only was its utility within mathematical physics known but its ontological status was clarified as well.

So here we have something whose non–existence was as firmly established as anything in logic, and yet it now exists. How did this miraculous birth take place? We all know the answer: by context extension. Within the context of real numbers, the square root doesn't exist; within the context of pairs of real numbers together with an appropriate algebraic structure, the square root exists. Mathematics is context sensitive. New contexts evolve; what was once impossible becomes possible. If one says, as one often does, that the possibility was always there but that we simply hadn't stumbled over it, one is oversimplifying the historical process and doing an injustice to our creative power.

G. J. Whitrow, a distinguished physicist who has written extensively about time in reference to events in the physical world, remarks that time is the mediator between the possible and the actual. To this we add that in the world of mathematical ideas, time is the mediator between the impossible and the possible.

If context is time–dependent, so are proof and verifiability. In the late eighteenth and early nineteenth centuries, a program of increased rigor in mathematical analysis emerged. Arguments which previously had seemed adequate were now thought to be shot through and through with logical gaps and mathematical nonsequiturs. The work of the great Swiss mathematician Leonhard Euler is a treasure of "invalid" procedures leading to valid results.[9] It would seem, then, that if mathematics is nothing more than a game of deduction, the rules of the game change even as the game is being played. Hedgehogs and flamingoes, indeed.

Mathematics is a growing, living fabric, a changing experience. In view of this change, how is it possible for it not to lead to chaos and unintelligibility as we work with it? To answer, we can compare it to the old cable cars of San Francisco. They make their ways up and down the hills of the city with open-air seats and a low speed. One is free to jump on or to jump off without too much fear for life and limb. The creative attention span of the working mathematician, limited in respect of time and knowledge, provides an experiential frame that is sufficiently steady to jump aboard. As David Bohm has pointed out, "The notion of something with an exhaustively specifiable and unvarying mode of being can be only an approximation and an abstraction from the infinite complexity of the changes taking place in the real process of becoming. Such an approximation and abstraction will be applicable for periods of time short enough so that no significant changes can take place in the basic properties and qualities defining the modes of being."[10]

Each attempt to view mathematics as existing outside of time and human society strips away a layer of meaning and exposes a desiccated kernel. The way in which the detemporalization is carried out is precisely by such a stripping process. Detemporalization leads to a naïve faith that formal manipulation may be productively and authoritatively invoked in any situation. It is often asserted that symbols have a life of their own, a power of creating that is almost magic: "symbols are wiser than we are." While there is truth to this—else why should man be the symbol-making creature he is?—temporalization prevents this tendency from being grossly abused.

A detemporalized mathematics cannot tell us what mathematics is, why mathematics is true, why it is beautiful, how it comes to be, or why anybody should care a fig about it. But if one places mathematics squarely within human time and experience, it becomes a warm and rich source of possible meanings and actions. Its ultimate mystery is never dispelled, yet it is exhibited as one of the prime creations of the human intellect.

Notes

1. See J.C.C. Smart, "Time," in *The Encyclopedia of Philosophy*, ed. Paul Edwards, 8 vols. (New York: Macmillan, 1967), 8:127.
2. Emile Meyerson, "The Elimination of Time in Classical Science," in Milic Čapek, ed., *The Concepts of Space and Time* (Boston: Reidel, 1976), pp. 255–64.
3. Norbert Wiener, Class Lecture at the Massachusetts Institute of Technology, 1949.
4. See, for example, "Comment on Meyerson's 'La Deduction Relativiste'," in Čapek, pp. 363–7.
5. See, for example, George Temple: "It must be admitted that most mathematicians are by nature Platonists who cheerfully, unreflectingly and habitually employ such loaded phrases as 'We assume that there exists . . .' or 'Therefore, there exits . . .' an entity with such and such characteristics. Challenged by the realist they would probably reply that since the truths of mathematics are absolute, universal and eternal it is hard indeed to deny them an existence independent of human intelligence" (*100 Years of Mathematics* [New York: Springer-Verlag, 1981], p. 4).
6. Smart, 8:127 (latter emphasis mine).
7. In Isaiah Berlin, *Vico and Herder* (New York: Viking, 1976).
8. Didier Norden, *Les Mathématiques Pures N'Existent Pas!* (Paris: Actes Sud, 1981).
9. See, for example. E.J. Barbeau, "Euler Subdues a Very Obstreperous Series," *American Mathematical Monthly* 86 (1979), 356–71, and Judith V. Grabiner, "Is Mathematical Truth Time-Dependent," *AMM* 81 (1974), 354–65. Grabiner cites the following shocker: "Here is how Leonhard Euler derived the infinite series for the cosine of an angle. He began with the identity $(\cos z + i \sin z)^n = \cos nz + i \sin nz$. He then expanded the left-hand side of the equation according to the binomial theorem. Taking the real part of that binomial expansion and equating it to $\cos nz$, he obtained

$$\cos nz = (\cos z)^n - \frac{n(n-1)}{2!}(\cos z)^{(n-2)}(\sin z)^2 + \frac{n(n-1)(n-2)(n-3)}{4!}(\cos z)^{(n-4)}(\sin z)^4 - \dots.$$

Let z be an infinitely small arc, and let n be infinitely large. Then:

$\cos z = 1$, $\sin z = z$, $n(n-1) = n^2$, $n(n-1)(n-2)(n-3) = n^4$, etc.

The equation now becomes recognizable:

$$\cos nz = 1 - \frac{n^2z^2}{2!} + \frac{n^4z^4}{4!} - \ldots.$$

But since z is infinitely small and n is infinitely large, Euler concludes that nz is a finite quantity. So let $nz = v$. The modern reader may be left slightly breathless; still, we have

$$\cos v = 1 - \frac{v^2}{2!} + \frac{v^4}{4!} - \ldots."$$

Which is a true equation.

10. David Bohm, *Causality and Chance in Modern Physics* (New York: Harper and Row, 1957).

Further Readings. See Bibliography.

S.G.F. Brandon; M. Capek; F.M. Cornford; A.R. Caponegri; P.J. Davis and R. Hersh; M.L. von Franz; J.T. Fraser; A. Koyré; D. Park (1980); D. Park (1985); C.M. Sherover; S. Weinberg; V.F. Weisskopf; G.J. Whitrow; R.L. Wilder

Non-Euclidean Geometry and Ethical Relativism

IT IS SOMETIMES asserted that contemporary intellectual thought is dominated by five isms: scientism, relativism, materialism, evolutionism and environmentalism.* Scientism asserts that science is the final arbiter of truth and values. Relativism asserts that no set of morals can be established scientifically, and hence all are equally valid. Materialism asserts that everything in the universe can be reduced to material objects and their interactions. Evolutionism, that everything evolves and hence that people have no special position in the universe. Environmentalism says that people are what their environment makes them.

It is also maintained that these five views have contributed greatly to the loss of meaning that is such a pervasive experience in our times.

The following dialogue with Professor Joan Richards deals with relativism. It explores the extent to which the rise of relativism was brought about by a crucial development in pure mathematics: the discovery of non–Euclidean geometry. Professor Richards teaches the History of Science at Brown University.

Non–Euclidean Geometry and Ethical Relativism

PJD: The role that mathematics plays in science and technology is pretty clear to most people, perhaps even the role that mathematics plays in business. Though people don't think of these roles consciously, they're aware of them. What they are not aware of is the role that mathematics has played over the centuries with respect to religion and ethics. The books on the history of mathematics written in the past hundred years hardly mention these topics. In 1973 a brilliant book was

* Eccles and Robinson, *The Wonder of Being Human*, p. 4.

Joan Richards.

published by Edward Purcell called *The Crisis of Democratic Theory*.[1] Purcell tries to uncover the relationship between the discovery of non–Euclidean geometry[2] and the "ethical relativism" that followed. A good place to start is a quotation that I found which comes from Walter Pater.[3] Walter Pater, in 1866, at the age of 26, wrote:

> Modern thought is distinguished from ancient by its cultivation of the relative spirit in place of the absolute. Ancient philosophy sought to wrap every object in an eternal outline, to fix thought in a necessary formula, and the varieties of life by kinds or genera. To the modern spirit nothing is, or can be, rightly known except relatively, and under conditions.

Pater was not a scientist; he was a literary critic and aesthete, and yet it struck him that he was living in an age of relativism. This was only a few years after the publication of the *Origin of Species*. What was going on in England at the time that might have elicited these words from Pater?

JR: The publication of *The Origin of the Species* is generally taken as a turning point in British intellectual life, but I don't think you find such turning points unless the society is ready to turn. In Britain in the late nineteenth century, major changes took place in people's perceptions of the nature of reality. Darwin's book serves as a very useful touchstone for understanding these changes, though I don't think the changes should be specifically attached to the book. The change which Pater talks about is not so much between ancient philosophy and the modern outlook: it's a description of the way a whole group of people in Pater's generation reacted against what they perceived to be an authoritarian interpretation of the nature of knowledge, which had been promulgated in the generation before them.

PJD: Do you think that Pater might have known about non–Euclidean geometry? Is it possible?

JR: No.

PJD: It's clear that he knew about Darwinism.

JR: It's clear that he knew. The Darwinian theory sparked a whole series of discussions which started at the level of scientific journalism, discussions between people like Louis Agassiz[4] and Thomas Huxley[5] and so forth, and very quickly were seen as relevant throughout the culture. It certainly would have been discussed by aesthetes like Pater.

PJD: Do you think that Darwin would have known about non– Euclidean geometry?

JR: No.

PJD: Nonetheless, both these theories, non–Euclidean geometry and evolution, are manifestations of—how did you put it—the readiness of the age to think about things in a different way than it had.

JR: Right. The interesting thing about non–Euclidean geometry is that the mathematical theory can be interpreted in a variety of different ways. In Britain, when I looked at the mathematical literature of about 1865, I found that Arthur Cayley[6] knew of the work of Lobachevsky,[7] which he had read about in Crelle's Journal,[8] and that a relatively unknown mathematician named Kelland in Ireland also knew of Lobachevsky's work. Neither of these men saw it as in any way relevant or important to larger issues of the nature of knowledge, of epistemology, ethics. They saw it as a minor aberration in mathematics; a curiosity. By 1869, only four years later, J.J.

Sylvester[9] was addressing the British Association for the Advancement of Science, and presenting non–Euclidean geometry in the way that it began to be understood in the 1870s. Here it's presented entirely differently—as a massive threat to the authoritarian approach to knowledge—but the theory of non–Euclidean geometry must be placed under a suitable lens to make it act in this way.

PJD: So the theory fostered a crisis by the 1870s.

JR: I'm not sure whether the theory fostered the crisis, or whether the interpretation of the theory contributed to, or was an integral part of, the crisis.

PJD: Was this crisis localized to Great Britain, or was it on the continent also?

JR: It was also on the continent, but in a somewhat different context. In Great Britain the connection between science, ethics, and religion had been strongly maintained through the early nineteenth century. In France it had not been, and in Germany there was a connection between mathematics and philosophy, very academically defined, but not in the broad cultural sense of an integrated society that one finds in the case of Britain.

PJD: In England, science, ethics, and religion had not been separated, so the crisis would have been broader there, and would have affected intellectual life much more. For example, the young Bertrand Russell[10] might have experienced it.

JR: Yes, definitely. The discussions of non–Euclidean geometry in England took place not just in the professional mathematical journals like the ones in Cambridge or Dublin, nor just in the British Association for the Advancement of Science and other large scientific organizations, nor just in the universities, but also in the pages of magazines like the Fortnightly Review, Contemporary Review, The Atheneum, the Dublin Review, a Catholic newspaper. The entire intellectual community was permeated with the concern, the implications, the meaning of non–Euclidean geometry.

PJD: What were the philosophical or ethical consequences when the crisis came to the fore, after, say, the 1870s and 1880s?

JR: Well, my knowledge of the history of ethics is not that great, but I think first of all it had consequences for the direction British mathematics took. In an attempt to deal with some of the specific epistemological questions which non–Euclidean geometry was raising, British mathematicians reinterpreted their perceptions of the nature of geometry and followed out rather idiosyncratic ideas. I should step back for a minute and talk about the epistemological assumptions on which the pre–non–Euclidean world view was based. The major assumption was that all knowledge was unitary. If one found a kind of certainty in geometry, then one could hope for the same kind of certainty in physics, in biology, in ethics, or in religion. All this knowledge stood on the same plane. Non–Euclidean geometry was seen as undermining the exemplar of perfect knowledge, classical geometry, and one very strong reaction was to divide knowledge, and to admit that geometry no longer held its position at the top of a unitary hierarchy. At the same time the study of ethics and psychology was split from the sciences. A more and more limited view emerged of what science was able to do, and people did not want to translate that limited view to what ethics was able to do.

PJD: Let's back up a bit. If you wanted to explain to the average person the distinction between non–Euclidean geometry and Euclidean geometry, how would you do it in a way which would explain the germination of the crisis?

JR: My explanation would be very non–mathematical. I would say that Euclidean geometry has stood for a long time as the exemplar of a kind of knowledge which was absolute and true. Particularly in British writings, you could find statements like: It is as certain that God exists as it is that the sum of the angles of a triangle is equal to 180°. This was knowledge which was not merely mathematically consistent but was both mathematically true and descriptive of reality. Now, that kind of knowledge was characterized by the inconceivability of its opposite. If you truly understood what a triangle was, then this seemingly inessential fact about a triangle became a necessary one. A human conception, a subjective conception, in this case a triangle, completely fit with a reality that we all experience. Now non–Euclidean geometry can be regarded as the notion that the sum of the angles of a triangle might not be 180°. All of a sudden, the whole superstructure of ethics, religion, and hope for finding true knowledge in the sciences collapses. That hope had rested on the belief that there was already at least one piece of such knowledge, but, now, it was no longer certain that the sum of the angles of a triangle was 180°, and so it was no longer certain that God existed.

PJD: Purcell says that this collapse was one of the streams that fed into what he calls "scientific naturalism." He says that scientific naturalism can be summed up in four points: that absolute rational principles do not explain the universe, that there are no *a priori* truths, that metaphysics is a cover–up for ignorance, and that only concrete scientific experimentation and investigation yields true knowledge. He says further that after Darwinism and non–Euclidean geometry came to the fore, there was a counterattack by the liberal church, but in the end scientific naturalism prevailed. One saw it in the new anthropology that was then emerging at the turn of the century. This new anthropology examined many different cultures and attributed to each culture its own validity and integrity. This he contrasts to the spirit of the Mission, which perceived non–Christian cultures as being in error and tried to convert them to Christianity. He says that by 1900 sociologists influenced by all this saw that change was natural and constant, that order was accidental, that process was non–teleological, behavior was adaptive, that values were those of the experimental laboratory, and that absolutes of any kind were superstitious and meaningless.

Would you agree that non–Euclidean geometry played a considerable role in the formation of these new attitudes?

JR: It's difficult to place it in a causal scheme. The changes in the view of the world that came about in the late nineteenth century were not caused by a single event or a single discovery or a single theory. The entire matrix or the entire world view, if you like, was shifting. That's why I brought up the example of Cayley and Kelland looking at non–Euclidean geometry before the new epistemological context burst on the scene in the seventies, and seeing in it nothing at all except mathematical detail. Within a certain context non–Euclidean geometry became a very powerful weapon. Outside of that context it might well have been overlooked, lost all interest, been ignored as it was ignored from the early 1830s until the 1870s. We can't evade the question of why non–Euclidean geometry was virtually ignored from 1830 to 1870. The time has to be ripe.

PJD: Classic Greek geometers were very brilliant, and they investigated many things. Since non–Euclidean geometry actually lay close to the surface, why wasn't it discovered hundreds of years earlier? Could you argue that what prevented it from emerging was the desire for an authoritarian basis for knowledge? Had Euclid been wrapped in the mantle of Authority?

JR: I don't think so. The books of Euclid mean different things to different people. There had been sporadic consideration of the problem of the parallel postulate. Then, starting in the early eighteenth century, there was a sharp acceleration of interest in asking what is the nature of Euclidean truth. The context of that acceleration was the aftermath of the new epistemologies that came out of the scientific revolution of the seventeenth century. In that context, it became very important to ask how man comes to know, especially how he comes to know science and mathematics, and the nature of that kind of knowledge. In that context, and only in that context, non–Euclidean geometry was, first, generated, and, second, interpreted.

PJD: Of course in today's mathematics the spirit of non–Euclidean geometry is everywhere. I mean by this the idea that mathematics consists of deductive structures in which deduction moves from axioms to conclusions. The axioms are whatever we choose; they are simply playthings. This, you might say, is the spirit of modern mathematics.

JR: But all this would be the Euclidean spirit, too, except for your own throwaway statement that axioms are merely playthings. It's that statement which neither the Euclidean nor the non –Euclidean geometers believed—that axioms are merely playthings. That point of view has been generated in response to some of the very fundamental challenges which non–Euclidean geometry posed.

PJD: Perhaps the discovery of non–Euclidean geometry forced the formalistic idea that axioms are nothing but playthings, they are simply starting points, and that the process of deduction that takes place in the mind or on paper may not have anything to do with what's going on in the outer world.

JR: It did, in one sense, force the idea, but I don't think it necessarily *had* to force the idea. Do you see the distinction I'm making? In the context in which it was introduced, it did force that, but, had it been introduced in a different context, it might have forced something totally different, or might have had no interest at all.

PJD: Again you're saying that non–Euclidean geometry, Darwinian ideas, relativist ethics and so on—are all part of a larger picture, in which all of them fit together each playing its role. The position of Eric Temple Bell[11] in the early 1930s is very interesting, and I'd like to describe it. Bell was a distinguished mathematician and a historian of mathematics. He was also a bit of a romancer and wrote science fiction novels. In 1934 he wrote a remarkable book called *The Search for Truth.* It now seems very much out of date, but I think that Bell spoke for the mathematical establishment of the 20s and 30s. Here is how I would summarize his book:

1. Mathematics is a tool created by the mind.
2. It has no relation with metaphysical or theological absolutes.
3. "Certainty has vanished and there is no hope of its return" (That's a direct quote.)
4. Mathematics cannot establish truth.
5. Mathematics has contributed four great landmarks to the history of ideas.

The first is measurement. Bell places this around 4000 B.C. The second is the notion of proof, and this he places around 500 B.C. He also says parenthetically that proofs are the chains that bound human reason for 2300 years. The third great breakthrough was non–Euclidean geometry in 1826, and the fourth was the then recently discovered multi–valued logics,[12] out of which he expected tremendous new advances in mathematics to arise. It's now 50 years or more since the discovery of multi–valued logics, and they seem to have been a rather sterile discovery. But certainly the existence of authority and the overthrowing of authority are pretty clearly on E.T. Bell's mind.

JR: I think it was clearly on the minds of people thinking about non–Euclidean geometry in the late nineteenth century. In my studies of the mathematical community of England, I find that the response was not a whoop of joy at the overthrowing of authority but, rather, an attempt to keep the authority there.

PJD: Do you agree with me that this position of Bell's is a little bit *passé*?

JR: Yes—and the audience that he's defining is an audience that is now gone. When he says, for example, there is no truth to be found in mathematics, the meaning that he attaches to truth is a meaning that not many people would now think of applying to mathematics.

PJD: One *can* establish truth in mathematics today, but the notion of truth has changed.

JR: Right.

PJD: And it's changed in response to these various crises.

JR: Whether in response to crises or whether it has just evolved, it has certainly changed. When someone talks about mathematical truth now they mean something utterly different than what they meant in 1860. Here is an interesting anecdote. Professor P. of the Math Department is sitting in on one of my courses. He is a very good student. He is very nice, very polite, and most of what I say is absolute anathema to him. A great deal of it has to do with the fact that when we say "truth," we talk right past each other. I use the word in an old–fashioned sense when I deal with the history of mathematics. He is a thoroughly well–trained, 20th–century mathematician who would never defile the name of truth in this way.

PJD: You would say that P. is a relativist?

JR: I'm in no position to go into P.'s views of truth. My point is that this particular word has radically changed its meaning. I feel that E.T. Bell is dated, not just because he puts multi–valued logics so high but because the audience Bell is speaking to is clearly not Professor P. Yet P. would be his current audience. It's that kind of datedness I feel.

PJD: Let's go back to the notion that the axioms are toys. I once was teaching a general undergraduate class, and I said that. A girl in the class literally screamed with non–acceptance. Now, where did she get that non–acceptance? Where does it come from? After all, she is a daughter of our age.

JR: Well, I think we get it from studying geometry in high school.

For example, there is a theorem that the sum of the angles in a triangle is 180 degrees. Of course, this is supposed to be proved from the axioms, but we all know that, despite what the teacher may say, despite the fact that this teacher will never accept it as proof, the fact is that you can draw 75 different triangles, measure the angles, and, in every single one of them, the angles will add up to 180 degrees. Correct mathematical statements about a triangle are true, not just in consequence of the abstract ideas from which they are generated, but also because they will be true if you draw a triangle and check. The demand to separate the axiomatic mathematical formulation from our spatial experience is a wrench for us all. When I teach a course on space, many of the students will mouth formalist platitudes, but, when I push them, they don't believe them at all. They see the mathematical theory and physical space as inextricably linked. They see Euclidean geometry as objective truth.

Professor P. finds it very hard to talk to such students. On the other hand, those students who come through the Math Department don't see it that way.

PJD: Well, there's a certain amount (if you'll excuse the expression) of brainwashing that goes on in the process of becoming a mathematician, and the brainwashing is at the metaphysical level.

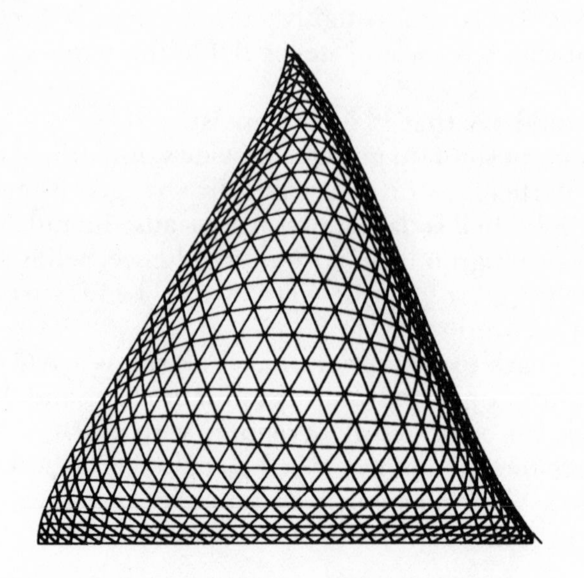

JR: Yes. I agree.

PJD: So we produce a generation of mathematicians who think along these lines, for whom mathematics is largely a game of deduction. The axioms are arbitrary. Some may be more interesting, some may be less interesting, some may be more productive, some may be less productive. 2 and 2 isn't necessarily 4; it depends on the particular system you're talking about. It's 4 in some systems; it's other things in other systems. This permeates all of today's mathematics.

JR: Yes, but don't blame the mathematician. A similar brainwashing permeates any specialized department in the university. This is one of the things that completely separates us from the system of knowledge and education which prevailed in Britain in the late nineteenth century.

PJD: One of the major points of Purcell's book is that he wants to show how ethical relativistic values foundered on the rocks of history in the 1930s. The swing of his argument is something like this. The attitude of historians in 1830, at least in the English–speaking world, was that of Macaulay: liberal; belief in progress and that England had achieved it. The belief of sociology, the attitude of law, was justification of the establishment. Throughout the 1800s the democratic ideal was a widely accepted axiom. Democracy was regarded as rationally and morally the best possible form of government. Religious faith, tradition, moderate rationalism going back to the Declaration of Independence, and concrete experience all testified to its vitality and the certainty of its ideals. This was the position in the early 1800s. Then came non–Euclidean geometry, Darwinism, and evolution. Scientific naturalism came in, and the new anthropology and the new sociology.

By 1900 William Graham Sumner[13] was able to assert that there is no natural law; there are no natural rights, and there is nothing which is *a priori*. "The only natural right is the right to struggle for survival." Now this comes into conflict with democratic beliefs, because if the only law of life is the struggle for survival, then the humanitarian ethics of democracy are an unwarranted handicap in that struggle. In Purcell's opinion, there began to develop a conflict between two ideals: the first is that of scientific objectivity, and the second is the moral ideal of a humane and truly democratic social order based on the toleration of opposing views. I think that today we are still in the midst of this conflict. One sees it on every level: the controversy over creationism versus evolution in the schools; the argument that any correlations between race and intelligence had better be left uninvestigated. From your point of view as a historian of mathematical ideas, how would you describe this conflict?

JR: By putting it in the context I developed before, of what has happened historically to perceptions of the nature of truth, and of what science has to say about ethical, religious, and social problems. In the early nineteenth century, science was seen as a study which could attain absolute truth. Then it was the highest ideal towards which all other investigations were striving. As scientific naturalists tried to undermine authoritarianism in science by pointing out the limitations of scientific knowledge, they also undermined the authority of scientific knowledge itself. The scientific naturalists said we cannot know what is absolutely true; only what is true within certain limits and relative to certain assumptions. Therefore we must give up our hope for absolute truth in ethics.

A more extreme response would be to say: if science is not true in the way it was claimed to be before, then we may ignore the scientists and pursue our own ideas wherever we want to follow them.

That reaction is an extreme in one direction. It's a complete rejection of science as a norm of truth, an attempt to knock science off its pedestal and to seek non–scientific approaches to human problems. At the other extreme, there are those who hold that in all fields, whether in ethics, in religion, or in any other area, there can be no human knowledge except what science can give us. This position leaves us at the mercy of such natural laws as science can find for us. This would be the view of someone like Sumner, who says there is no natural law for society other than the survival of the fittest. One alternative conclusion to Sumner's might be: perhaps natural law is not the kind of law that ought to govern society. There might be other kinds of law, aside from the survival of the fittest; human laws which operate within society.

PJD: The modern spirit that Pater wrote about, the spirit of relativistic thinking, emerges in a fierce, sudden way. You have implied that it's partly a generational phenomenon. Purcell's explanations are in terms of the history of ideas: what it is that people think explains what people do. There must be more to it than generational. Is there, for example, a Marxist explanation?

JR: Certainly. A Marxist explanation would say that there was a class struggle going on within mathematics education in England. The upper classes were being educated at Oxford and Cambridge. Mathematics was studied there not for its own sake, nor for the sake of its use in science, but in order to learn the nature of absolute truth. On the other hand, there was a growing demand for education designed for middle class students. This meant useful education, education which could be used to promote British manufacture and British trade. For the middle class, the study of geometry was not seen mainly as the study of absolute truth, but rather as a study which was useful to engineers, what we should call applied mathematics. A useful curriculum was being demanded by Huxley and by Clifford[14], who was then teaching mathematics at University College, a much more middle class place. The interpretation given to geometry there was much more scientific, by which I mean inductive, empirical. This fits in with the pragmatic, utilitarian view of the importance of human knowledge. During the 1860s the middle classes were gaining in strength and in vociferousness. They were attacking the ancient bastions of upper class education. You can see this most clearly in the attack on Euclid which was mounted in the 1860s and 70s. A number of England's teachers of mathematics tried to get rid of Euclid as a textbook, saying that it wasn't appropriate for young boys, it wasn't developmentally sensible, it wasn't psychologically appropriate, and so forth. They tried to change the syllabi and the criteria by which geometrical knowledge was tested and to turn it away from memorization of Euclid's theorems and Euclid's proofs toward what they saw as a truer and deeper understanding of geometry. This particular battle they lost, at least at the Cambridge level. Euclid remained the key to entering Cambridge throughout the late 19th Century. The thing to notice is that Euclid itself was the prerequisite, not geometry at large—even if you had a knowledge of geometry and could solve geometrical problems, you still might find yourself unable to matriculate in mathematics at Cambridge if you could not spout the specific theorems and proofs that Euclid gave.

PJD: What you're saying is that Euclid was part of the classical tradition, the classical tradition was part of a class tradition, and this was called into question.

JR: Yes.

PJD: Let's try to wrap up our discussion. On balance, to what extent did non–Euclidean geometry contribute to the new relativistic spirit of the age? Conversely, to what extent did this new spirit contribute to a new interpretation of the meaning of mathematics?

JR: To see non–Euclidean geometry as having relevance to relativism, you need a receptive intellectual community. Until the appearance of a receptive community, the issue was ignored. Lobatchewsky and Bolyai published in the 30s, but nothing happened at all 'til the 60s and 70s. When it did happen, first in Germany, then in England and France, it happened very strongly. People discovered Lobatchewsky and Bolyai; they discovered Riemann; Helmholtz did his work. It was a sudden outburst that needed a receptive intellectual community able to perceive the mathematics as relevant.

Notes

1. Edward A. Purcell, *The Crisis of Democratic Theory*, Univ. of Kentucky Press 1973.
2. Non–Euclidean Geometry. A geometry that is not based upon the axioms of Euclid. More specifically, a geometry that rejects Euclid's parallel axiom.
3. Walter Pater, 1839–1894, English essayist and critic. The statement reproduced can be found in his essay on Coleridge in "Appreciations," Macmillan, London, 1910.
4. Louis Agassiz, 1807–1873, Swiss–born naturalist at Harvard. Agassiz rejected natural selection.
5. Thomas H. Huxley, 1825–1895, British biologist who advocated Darwinism.
6. Arthur Cayley, 1821–1895, British mathematician.
7. Nikolai I. Lobachevsky, 1793–1856. Russian mathematician. Co–discoverer with János Bolyai, Hungarian, 1802–1860, of non–Euclidean geometry.
8. Crelle's Journal. One of the most important mathematical journals of the 1800s.

9. J.J. Sylvester, 1814–1897. British mathematician who taught at the University of Virginia and at Johns Hopkins.
10. Bertrand Russell, 1872–1970, British mathematician, philosopher, and social activist.
11. Eric Temple Bell, 1883–1960, American mathematician and historian of mathematics.
12. Multi-valued logic. A formal logic in which statements can have truth status other than true or false.
13. William Graham Sumner, 1840–1910, Professor of Economics and Social Sciences at Yale.
14. William K. Clifford, 1845–1874, British mathematician and scientist.

Further Readings. See Bibliography

J. Barzun; J. Ladd; J. Richards; I. Toth

The Unreasonable Effectiveness of Computers. Are We Hooked?

PJD: A generation ago, the famous physicist Eugene P. Wigner wrote a much quoted article entitled "On the Unreasonable Effectiveness of Mathematics in the Natural Sciences." The basic question was: why is it that mathematics works so very well in physics? I assume that we're talking about something different here. We're not talking about why computers work in physics.

CS: No, we're talking about why they work in our society, in western civilization. Why they're becoming so ubiquitous.

PJD: Could you conceive of a society that says: "To hell with them? We don't need them?"

CS: Our society is one that just went right into them—it was a love affair from the start. In 1950, the best marketing minds at IBM said that there might conceivably be room for a hundred large scale computers in the world. By large scale they meant what we would now call a mini computer (a million bits). And, of course, now they're everywhere. They're proliferating faster than anything. Obviously, some kind of need is being felt for them. They must be of some plausible use, otherwise there wouldn't be so many of them.

PJD: Do you see any civilizations on earth today that are resisting them in any way?

CS: I don't think it's a matter of resistance. There are some pockets of civilization—humanities departments in universities for a long time thought that computers were a work of the devil, but a minor devil so that they could easily be ignored. They had managed to ignore science, so why couldn't they ignore computers? Even they are being taken in by computers, leaping onto the bandwagon with text processing and office mechanization and things like that.

PJD: Would you say that there are pockets of computer Luddites?

CS: Of course there are huge stretches of people who simply don't care. Let's say the Aborigines of Australia. I doubt that computers have really gotten a large hold on them.* They just don't care. Now there's something about our civilization that really cares about computers or about what they can do.

PJD: Why do we love them so much? What are they doing for us?

CS: Everything. That's what's so remarkable about it. If it were only in a few areas of competence, say accounting or physics or specialized things like that, or even fairly general things but only in a few areas, I'd say it was an accident. It's a matter of matching the man or the job to the machine, and this was a nice meld. But I think the spread of computers is literally remarkable. In the United States just about everything is now considered fair game for computers. In factories, in homes, in offices, everywhere.

PJD: Isn't it a case of the function seeking the form, in that the computer form is available? We look around desperately for functions that can fit into this form. "I've got the answer, what's the question?"

CS: No, because there are all kinds of other forms around. If you look at what computers do, you see that what they're good at is dumb, repetitive tasks (in spite of the AI people). Now, they're built in a dumb, repetitive way. The economics of building computers in the last 30 years has been in the direction of simpler, more standardized, minifunctions. Less functionality per component, but much more standardized, and many more components. Lots of very standardized components rather than fewer, but less standardized, components with more functions. There are sound engineering reasons for this, but it could well have gone the other way. Computers do very straightforward things best. There is so much in our civilization that is amenable to a machine that just does dull repetitive tasks.

* Having just returned from an international conference on mathematical education held in Adelaide, South Australia, at which quite a few aboriginal teachers were present, I think CS should look for a better example.

PJD: I assume that the cosmos or the solar system with its built-in periodicities of the day, the month or the year forces periodicities on our human activities. There are things that we have to do daily, monthly, and so on, over and over again. We have to send out bills monthly.

CS: There's another kind of repetitive activity that comes from the spread of consumerism. We want to make an automobile, and then two, four, eight automobiles, and in this way we get a different kind of repetition.

PJD: Back to the topic of resistance. A number of years ago Kurt Vonnegut wrote a book that was very popular in its day, and the title was *The Player Piano*. As I understand it, the implication of the title was that the player piano was a mechanism which supercedes the human. It's also a metaphor for the computerization of the world. The interesting thing is that the player piano reached its height in the 20s or maybe even earlier, and you as a piano player probably understand the psychology of this dynamic process. Now why would you say that the player piano peaked?

CS: First of all the piano as a purveyor of popular music or music in general peaked around then. They managed to perfect the radio and the phonograph.

PJD: So you're saying the radio and the phonograph are in fact substitutes for the player piano?

CS: Yes, and better, if you think of the range of music that can be reproduced well on them.

PJD: As a piano player would you want to walk into a party and push the button to set the player piano in motion? Is there a tendency back to the "do it yourself" in music? It can be argued that the computer's doing all the dull, repetitive tasks has freed us up. Do you see an indication of this in the musical field?

CS: Yes. I'm not quite sure where it comes from, but I've certainly noticed over the last five years that there's been much more of a push in bars and restaurants to have live music, whether it's a small combo or a single piano player. There's much more live music available now than formerly.

PJD: Is it conceivable that, in the wider field where electronic computers are applicable, one might see a counter push toward doing things by hand for the simple joy of it?

CS: Yes, but it would be on the periphery. You don't get a counter push until you have a push, and the push is coming from the introduction of computers all over, and the subject of the discussion today is why have people felt impelled to go to all the trouble of bringing in computers in the first place. They got along perfectly well without them 30 years ago.

PJD: I see a widespread tendency to computerize wherever possible. For example, computers are going into libraries now, and one has the feeling that librarians can't keep their heads high unless they work for a library which is computerized. You can multiply this in many fields. Is it clear that the computerized library, for example, is going to give us better service than the old fashioned kind?

CS: Again, it allows a higher volume of more standardized processing. Does the library you're thinking of have computerized retrieval, or is it just computerized as far as checking the books in and out is concerned?

PJD: Well, the system doesn't exactly send a mechanical finger into the stacks to retrieve the books and deliver them.

CS: Then you're talking about the computerization of the card catalogue?

PJD: At the moment, I'm talking about the computerization of the withdrawal procedure.

CS: There I suppose it's because computers are getting cheaper and cheaper, and people are getting more and more expensive. So they said okay, we'll go for a quarter or a half or two million dollars at the beginning, but we'll make it back. You know librarians are unionized now, even the part–time helpers are, and I guess the library administration foresees that they would be having trouble within a few years. I guess they figure computers don't strike. The worst you have to worry about is having the power cut off.

PJD: One of the bad aspects of this, as far as I can make out, is that the libraries that are computerizing have lots and lots of money for computer systems and none for buying new books. So there's a perfect example of how the medium becomes the message.

CS: Okay, but what good would all the new books do you if all of the librarians were forever out on strike, and you couldn't get to the books?

PJD: So there's a trade off here.

CS: Yes, I'm now arguing on the computer side. There are arguments *for* the introduction of computers in that particular situation.

PJD: I suppose that the computer business has a tremendous momentum of its own. There are hundreds of thousands of people making a living thinking up new uses for computers. If you start thinking up new uses, you've got to push them on civilization, or compel civilization to adopt some of them.

CS: The pull has to be there, too. You cannot sell things to people who really don't want them. This is not a fad anymore, this is big business.

PJD: Let's compare computerization—I don't know if the analogy is fair—but let's compare it with the love affair with automobiles that started in the twenties. You see there the push and the pull. Certainly there was a psychological pull towards movement, speed, flexibility. The automobile gave us a lot of possibilities that weren't around in the horse and buggy days, and then the system started choking up. Could you put your finger on the pull in the computer area?

CS: I could certainly put my finger on the pull in the car area, because at various times I've been without a car for six months. It was extraordinarily inconvenient, especially psychologically inconvenient, because there would be places that would be trivial distances in time—not space—to get to. A distance 10 minutes by car is practically impossible to walk. Ten miles is hard to walk; it's especially hard to walk and then do something and then walk back. It's terribly hard if you're carrying bundles one way or both. You just don't do it. That's a difference in degree that's a difference in kind. Being without a car was like being bedridden.

It was like being kept in the house because the weather was so bad you couldn't go out and play. Not that I would have zinged off to all of these places that much, but it was the inability to do it that was oppressive.

With computers, the pull, certainly part of its glamour, is that it's the modern way. When you go into a hi–fi shop, and you look at the outer shell, the box, of the turntables and preamps, etc., both the shell and the panel are interesting. For a while there were dials, then slides, and now I believe the latest thing is extreme simplicity. You see a black plastic wall and it cues you (again it's done by computers, they have microprocessors built in). You press a query button, and things light up on the panels. There's nothing projected and nothing recessed; it's all a plastic or glass panel which is lit up with a color coded square to show the state that it's in. Something happens—it gets louder or you shift channels or something. It's amazing!

This is part of industrial design psychology. In the same way the industrial design psychology of computers is romantic, flashy; it's "with it."

PJD: So computers are simply an expression of the modern period?

CS: They are our symbol, our logo in a way. We adopt them, and they change us. Let's look at the office of the future, office automation. The office has remained the same since 1850. New machines have been introduced to do very specific tasks—calculators, typewriters, copiers, but they never changed the office flow very much. Computers change this flow. You cannot have a computer network and not really change the way the whole complement of people in an office relate to each other, or not change their status—because information and access to information are being traded here. Not everyone has access to the same information. The standard boss–secretary–file clerk relationship where the boss says to the secretary "Get me this file," and the secretary tells the file clerk and the file clerk has to go to a physical file and pull out the file—all that goes by the boards when you automate the office. That means that the people who know the keys and who can diddle the machine are now very important.

PJD: Come back to the economics of the thing. I have the feeling that the dynamism that the computers express these days is not entirely due to the economic advantage that they confer but is simply the way we have to do things at this moment in time. Why?

CS: Well, you mentioned glamour. Glamour is one reason. Another is cost projections of the future. The office is very labor intensive—people are going to cost money, and, what's more, people are unstandard. You're forever having to cajole them and fire them and hire them. Granted, systems programmers are terrible people, but, then again, personnel people are terrible people too.

PJD: It's harder to service a person than to service a machine that's down.

CS: Yes. Or whether it is or not, it's certainly perceived that way.

PJD: I don't quite understand the economics of the thing. Granted that labor is very expensive, and, if you computerize, you avoid certain labor costs, but then you have the back up computer industry where the people are now employed, and presumably they're making a salary there.

CS: Yes, but the great thing about software is that once you write it, you can replicate it forever and it costs nothing. The prototype is always very expensive.

PJD: You don't envision a civilization in which most people are unemployed?

CS: No, but I'm wondering what the implications of all this are. So far, the computer has created at least as many jobs as it's destroyed. Still, there are an awful lot of jobs that it's changed. I wonder if it hasn't contributed in large part to the lack of need for manual labor, for unskilled labor.

PJD: Let's focus on the word "effectiveness." In what sense is the computer unreasonably effective?

CS: There is a quote from Kingsley Amis' first book, *Lucky Jim*, where Jim meets a pretty blond girl with large breasts, and he wonders to himself why he likes breasts so much. He knows perfectly well why he likes breasts, thank you very much, the question was why he liked them *so* much. In the same way, I can certainly see (I've worked with computers for 20 years now) that they're good for a lot of things, a lot of specific tasks I wouldn't want to do any way but with a computer. What I find amazing is that they are so all pervasive in every aspect of our civilization. Some of it is because of what some people would call unreasonable governmental type demands. A pressure group leans on the U.S. government that some statistic ought to be gathered so that something is not going to be unfair or so that money gets apportioned in what seems to be a fair manner. So the demand gets passed into law and the word goes out—you've got to fill out this form quarterly. Everyone in a certain line of business, or every recipient of government money, or something like that, has to fill out this form and provide this information.

Well, one form, who cares—you fill it out once a year. You multiply this though; there are lots of forms. There's just so much that a particular business does, and if you have to keep on filling out form after form after form about it, then at some point it becomes cheaper to say, all right, we have a finite amount of input. That is all the information about our business. Let's computerize it. Let's put it into some kind of information retrieval system. Let us attach to that information retrieval system a generalized report writer, a report filler outer would be a better word. And then I don't care what the bastards ask for. They can ask me for anything and I will simply go ticka ticka ticka in some reasonably easy manner. And then whenever they want that information—monthly, weekly, daily, yearly—it will all be in the standard system. When I come in in the morning I'll say "Give me today's stupid forms," and the computer will go ticka ticka ticka, because it will have all the information. It will have pooled all the information that these forms tap off, and, bingo, it's done. In other words, it puts a ceiling over the amount of effort and expense to which you can conceivably be pushed.

PJD: It seems to me that this is an instance of the system's posing demands which are fulfillable by computer, because previous demands of that sort were filled by computer. What is that, Murphy's Law or something?

CS: Parkinson's Disease. It seems to me that part of the unreasonable effectiveness is due to this Parkinsonianism.

PJD: Do you see any computer saturation in the foreseeable future?

CS: There's got to be a time when falling costs will not cover the diseconomics of scale, when all the easy jobs will have been done and it will become harder and harder to introduce them into the jobs that have been left undone. Sure, then it'll slow down, except I cannot think when that will be. There are some fields in which computerization hasn't taken place yet, for example in the construction of an automobile, i.e., in the individual automobile. One sees computerized dashboards with tricky readouts, and one might think that that would progress for a while.

PJD: What about the little chips in VW's to handle ignition, timing, etc. And the self–diagnosis. When you get VW service these days, you drive it in, and they plug a tester in. The tester's microcomputer speaks to the engine's microcomputer, and one tells the other what's wrong with it.

CS: It's here. It's amazing to me, but it's here. What hasn't been automated terribly well is the manufacturing process, except in Japan, of course.

PJD: There's been a lot of hype lately about the future role of the computer in the household. I've not thought it convincing. Do you have any comment?

CS: I can't see it at all. All the instances I've seen have been strictly hobbyistic.

PJD: Since the effectiveness of the computer is unreasonable, we can't really explain it completely.

CS: No, we can't. I used unreasonable because it seems at first glance that here is a device which is good for a certain number of things, and it turns out to be good for many, many more things. It is invading areas that until as recently as five years ago, except in science fiction, seemed absolutely sacrosanct. Like cars. I've seen science fiction stories that always have the computer steering the car instead of what they actually do—diagnostics to say what's wrong with the car. I never thought that our civilization was capable of having large chunks of it standardized. No, that's not strong enough. I never really thought that our desire for a lack of trouble or minimization of trouble would have been worth the agony of going over to the crisp, clean, slick, no–nonsense standardization required by computer. People seem to be willing to change hundred–year–old routines in order to computerize all, or parts, of them. That really surprises me.

PJD: What is the implication of this for the psychology of the future?

CS: I should think that once having changed, people would not want to change again, but I find that hard to say because I've been working with computers all my life and think naturally in computer terms. I haven't really the faintest idea, and I don't think anyone else has thought what the unforeseen macroscopic psychological effects are going to be. Increased alienation? Perhaps less alienation. Will it free mankind for human interactions instead of having to do all the dull repetitive tasks? Who knows?

PJD: At the time of the gas shortage people tried to visualize a civilization without automobiles or with some kind of alternative transportation. We didn't succeed very well.

CS: Wait a minute—all you have to do is look back to 1930 when most people didn't have automobiles and things were reasonably livable.

PJD: That's true, but we aren't ready to go back to 1930.

CS: That's because we've built the shopping centers out of the cities and all the suburbs. Now we're hooked. It's very hard to go back. We could go back, though. We don't want to. Computers hook us. We can unplug an individual computer but not the whole computer civilization. For better or for worse, we are now hooked.

Further Readings. See Bibliography

J. Deken (1982, 1983); M. Dertouzos and J. Moses; C. Evans; E. Feigenbaum and P. McCorduck; P. McCorduck (1984); H. Pagels; H. Sackman

V

MATHEMATICS AND ETHICS

(Courtesy of the Lowndes Collection, John Hay Library, Brown University.)

Platonic Mathematics Meets Platonic Philosophy of Religion: An Ethical Metaphor

WHAT IS THE RELATION between mathematics and God? Of course, there are many people, in particular many mathematicians, who believe in mathematics but not in God. On the other hand, many more people, by far, believe in God but not in mathematics. Here we have already established a parallel between the two.

More seriously, it can be seen on brief reflection that the Platonist conception of mathematics (held by nearly all mathematicians nearly all the time) attributes to mathematics certain properties and features usually attributed only to God. This observation suggests that we pursue the God–mathematics correspondence a little further. We are by no means supposing that the two entities are the same, only that there is a certain parallelism between the two which might be interesting to follow up.

To do this, we call upon a great Neo–Platonist religious philosopher, Philo Judaeus of Alexandria (20 B.C.–50 A.D.), as edited and described by his outstanding recent expositor, the late Harry A. Wolfson. We will take Philo's description of God's properties and attributes, and see what happens if these statements are transformed by replacing "God" with "mathematics." In this way, we will create a metaphor, which might be termed a religious aspect of mathematics.

Let us begin with Spinoza. In his "Ethics," Spinoza asks himself, according to Wolfson,[1]

"What is the most fundamental assumption underlying the religious philosophies of Judaism, Christianity and Islam? . . . It is the belief that over and above and beyond the aggregate of things which make up this our physical universe, there is something unlike the universe . . .

Harry A. Wolfson. (*Photo by Irene Shwachman.*)

They all believed that, unlike the world which is dependent upon God, God is independent of the world. This independence of God may be expressed by the term separateness—separateness in the sense that the existence of God does not necessarily imply the existence of the world. For believing, as all of them did, that the world came into existence after it had not been in existence, they also believed that prior to the existence of the world, there was a God without a world. And believing, as some of them did, that some day the world will come to an end, they also believed that after the ultimate destruction of the world, there will be God, again, without a world. And since God was, and will be without a world, even now when the world exists, God's existence is independent of the world, separate from it, and apart from it."

Now we shall set up our metaphor, by substituting for the word "God" the word "mathematics" or "mathematical knowledge." If we make this substitution in the opinion just quoted, then one has a view of mathematics, commonly called Platonic, which is held, more or less, by many people. In this view, mathematics is independent of the world. It existed prior to and apart from the world, and if, some day, the world should come to an end, the existence of mathematics would go on. By the word "world" here is meant not only the planet Earth, but the whole physical cosmos. In this view, the task of the mathematician is to discover the mathematics that is already "there," to record it, arrange it, and discuss it. Mathematics is often applicable to the physical world, and these applications are a confirmation of its existence and its power, but such applications are just a bonus. The existence of mathematics is prior to and independent of them.

If this view seems hard to grasp, one might explain it by saying that one has, in God, an instance of something which is self–existing, self–knowing, and self–justifying, so that to assert the same for mathematics is a parallel assertion, not without precedent.

The relationship between God and mathematics has been discussed, admittedly in a casual way, by authors earlier than Philo. "All is number," said Pythagoras. "God is a Mathematician," is a modern formulation meaning that the way of the world is mathematical, that mathematics provides the key to the universe, that God, as the Prime Mathematician, set up the universe according to the principles of mathematics. This view may be slightly egocentric, perhaps, and not necessarily subscribed to by theologians. It is a view that is widely held today by physicists (who may or may not use the word "God") in order to answer the unanswerable question of why mathematics is such an effective tool in theoretical physics. It is the view which lies behind a great deal of the recent mathematizations of a variety of disciplines, history, sociology, psychology. The world *is* mathematical, and hence, to interpret it properly, one must use mathematics.

There is a related view, also of recent advocacy,[2] which turns this around somewhat and asserts that the "total science of intellectual order" is, automatically, mathematics. We have now arrived at a complete equivalence: what runs the world is mathematics. What is mathematics? What runs the world.

The ancient religious philosophers would assert that God is prior to mathematics (as to all else). He created it, and it existed prior to the physical universe, observed by him. This view now runs into a difficulty. One of the attributes of God is his omnipotence. "Omnipotence" has been defined variously. According to Philo,[3] four things are meant by this word.

"(a) God created the world out of nothing and implanted in it certain laws of nature by which it is governed.

(b) Before the creation of this world of ours, God, if He willed, could not have created it at all or could have created another kind of world governed by another kind of law.

(c) In this present world of ours, God can override the laws which He himself has implanted in the world and create what are called miracles.
(d) God, if He wills, can destroy this world and create in its stead a new heaven and a new earth."

Now God's omnipotence runs up against the notion of the permanence and the infallibility of mathematics. If God is truly omnipotent, He could create a universe in which mathematical and logical impossibilities exist, in which, for example, $2 + 2 \neq 4$: He could completely turn around the content of mathematics. On this point, philosophical opinion is split, one opinion being that God's omnipotence includes the power to

> bring it about that it should not follow from the nature of a triangle that its angles should be equal to two right angles.[4]

The second opinion denies this. It sets a limit on God's power and therefore seems to separate mathematics both from God and the world.

Within the first opinion, God is placed at the pinnacle of power and possibility, below him are the mathematical laws, and below the mathematical laws are the physical laws.

To abrogate the physical laws as in the scriptural miracles did not worry the religious philosophers as much as the abrogation of the mathematical laws. Thus, mathematics is accorded a distinguished position, and the possibility of its eternal truths being abrogated, even by an omnipotent God, is disturbing.

The scriptural philosophers try to effect a reconciliation between these contradictory positions. This reconciliation, according to Wolfson,[5] goes along the following lines:

> It is not for the lack of power, he would argue, that God does not change impossibilities; it is rather out of wisdom and justice. God could have created another world in which these impossibilities would have been possibilities. He can also destroy this world and create a new world in which the impossibilities would become possibilities. But having by His wisdom created this world and implanted in it these laws, He would not change these laws except that it served a certain purpose. For God does not change the laws of nature in vain, nor does He, like a stage magician, perform miracles to amuse or to impress the spectators. Miracles are performed and laws of nature are changed by God only in His exercise of individual providence, for the purpose of preserving those who deserve to be preserved, or for the purpose of instructing those

> who deserve to be instructed. Now, in the wisdom of God, the world is so ordered that to attain that purpose of miracles there is only a need for a change of the physical laws of nature; there is no need for a change of the laws of thought or of the laws of mathematics. All miracles recorded in the scriptures from the creation of the world to the resurrection of Jesus are miracles which only involved a transgression of the physical laws of nature, for these miracles had a purpose. No conceivable purpose could be served in the world as it is presently constituted for a miraculous change in the laws of thought or in the laws of mathematics. When scriptural philosophers, therefore, say that God does not change, in this world as it is presently constituted, the law of contradiction or the geometrical proposition about the three angles of a triangle, it is not an indication of a lack of power; it is an indication of the fact that God uses His power in accordance with His wisdom and His goodness.

It is interesting that Spinoza selects as the mathematical truth to be abrogated by omnipotent God precisely that fact which was abrogated (or made optional) by mathematicians themselves without divine intervention, in their creation of non–Euclidean geometry.

The difficulty that the scriptural philosopher faced was that of confronting an omnipotent God with a Platonic view of mathematics. Mathematics is set up as a rival to God, but once the Platonic view of mathematics is abandoned the philosophical difficulty evaporates.

Insofar as Platonic mathematics is a rival of an eternal, all knowing, omnipotent God, it is not an accident that many people turn to mathematics, consciously or unconsciously, as a substitute for religion. There is a strong craving for permanence, for certainty in a chaotic world, and many people prefer to look for it within a mathematical or scientific rather than a religious context. They are, perhaps, not aware that underlying both mathematics and religion there must be a foundation of faith which the individual must himself supply.

Religious philosophy concerns itself, in part, with the question of how one arrives at a knowledge of God. If we substitute the word "mathematics" for "God" to work our metaphor, we arrive at another serious question: "How does one arrive at a knowledge of mathematics?" Let us refer, once again, to the opinions of Philo. According to Philo, knowledge of God may be arrived at in three ways: the way of imagination, the way of reason, and the way of revelation.[6]

"The way of imagination" is the view that God is a mere concept, a product of our imagination, that has been invented for certain utilitarian purposes — mainly to inspire reverence for civil law and order.

In our mathematical analogy, "the way of imagination" may be most closely compared with the philosophic position of formalism. This asserts that there is no inner content to our mathematical symbols and manipulations, that mathematics is a formal "game" played along certain lines because it is convenient for certain purposes. Philo says that "the way of imagination" is atheism. Correspondingly, we would say that the way of formalism provides an inadequate picture of mathematical activity.

"The way of reason," in religion, according to Philo, leads to the God of Plato, of Aristotle, and of the Stoics "who starting from a world in which they saw order and beauty and purpose found themselves compelled by reason to arrive at the existence of God."[7]

Now mathematics is usually held to be "the way of reason" par excellence. Deductive proof within formal axiomatic systems is the hallmark of mathematical sciences. It is the ideal goal of mathematical exposition and the certification of finality to our mathematical knowledge.

Philo believes that reason is not the only way of arriving at a knowledge of God, nor does reason lead to a true and full knowledge. In the mathematical field, as well, reason should not be thought of as the sole road to knowledge. Mathematics today pays great homage to deductive reasoning and accords it the power of issuing the final certification, but for making new mathematics there are other processes which are fully as important.

The third philonic mode of acquiring knowledge is *revelation*. Now, to Philo, revelation meant two things. There is firstly the historic revelation: a body of divine material transferred from God to Man at a definite time and place. In short, the Bible. Secondly, revelation meant progressive revelation through individual humans who are divinely inspired and through whom new truths are discovered and the real meaning of old truths is found.

In mathematics, revelation of the first type does not exist. There is no book or body of material which purports to be of divine origin and which serves as a point of departure for mathematical discovery.

On the other hand, revelation in the second of Philo's senses exists and is of great importance. The work of individual geniuses is recognised, revered, and studied assiduously. Mathematical intuition, the "inner light," has always been a key ingredient of mathematical life, along with real–world experiences, experimentation and deduction.

Let us pass now from the knowledge of God to his attributes. Here, analogy becomes more tenuous but still, I think, worth formulating.

Philo concentrates on three attributes, all having to do with God's infinitude. This is in contrast to the concept of a finite God which was then common in the Greek world.[8] The "attributes" of God are that He is infinite (a) in the sense of incomprehensibility; (b) in the sense of infinite goodness; (c) in the sense of omnipotence. We have already talked about the last, omnipotence. The incomprehensibility of God, attribute (a), signifies that it is impossible to compare God with anything else whatever. He is unique. He belongs to no class and no concept of Him can be formed. (Compare this with the paradoxes of set formation in naïve set theory!)

Continuing our parallel wherein we have substituted "mathematics" for "God," this corresponds to the doctrine that mathematics is basically incomprehensible. This comes close to Bertrand Russell's joke about mathematics, that it is the subject in which one never knows what one is talking about nor whether what it says is true. It also is close to the formalist position that it matters not what the ultimate reality is but only whether we manipulate our symbols properly.

The second attribute, "God is of infinite goodness," meant several things to Philo. It implied that God acts freely and purposefully and that he exercises his providence over individuals. That is, he is concerned not only with great cosmic matters but also with the smallest detail of the universe. "The sparrow doth not fall, but that he knows it."

Continuing with our parallel: mathematics is purposive. That is, it is not a meaningless game, played freely and mindlessly. It has meaning, utility, implication. It is concerned with small detail; a mathematical structure is built up from myriads of tiny, seemingly inconsequential, items. Yet the failure of the most inconsequential member, it is thought, threatens the stability of the whole enterprise.

We come now to the interpretation of the expression "infinite goodness" in the moral sense. In the religious sphere this raises a philosophical problem of the first magnitude. If God is good, and if the world derives solely from Him, how comes it that evil exists? By evil is meant both physical and moral evil. Philo calls on all his resources in his attempt to reconcile the goodness and the omnipotence of God. He uses philosophical arguments, and he uses traditional folk arguments; despite all arguments, the problem is very large, and he must have recourse, finally, to the principal of the incomprehensibility of God.

Now, if mathematics is eternal, and true, and good in the sense of being unflawed and expressing deep truths about the universe, how comes it that out of mathematics, or in conjunction with mathematics, can flow evil, both physical and moral? This is not an empty or absurd question. From Archimedes (200 B.C.), mathematician and scientist, who applied his knowledge of optics and mechanics to design catapults and burning glasses, to J. Robert Oppenheimer, director of the development and construction of the atomic bomb, whose agonies of guilt are recorded in his autobiography, there is a constant record of mathematics at the service of military construction and destruction. More generally, through its connection with computers and with applied statistics, there is hardly any noxious tendency in modern society, from mass advertising through cannibalistic uses of natural resources down to gambling casinos in Nevada and New Jersey, in which mathematics is not implicated as an accomplice.

How can we attempt to reconcile this dilemma? Even if we keep Plato's identification of mathematics with "The Good," we can get around the difficulty by changing the sense of the word "good." We can say that science—and likewise mathematics—is morally neutral. "It can be put to bad as well as good uses." The parallel statement in religion would be that God is neutral, or indifferent. If we do this, the meaning of the word "good" is transformed, stripped of ethical implications. Then mathematics may be the "total science of intellectual order," but the intellect has no heart.

Another attempt at reconciliation says that it is not mathematics as a whole, but only a part of it, which produces evil. Some people have called applied mathematics corrupt, while pure mathematics (pure in the sense of being disconnected from applications) is defended as morally pure. We suspect that no matter how subtly the decision is made, the part which remains and is now called "good" will be perceived in future generations to be the generator of as much evil as good.

The reconciliation of this dilemma cannot be achieved by exclusion or excision. If it is attempted along the lines of Philo Judaeus, we shall be forced into the Philonic "cop–out", bewildering but perhaps realistic, that, while mathematics is comprehensible, its ways, as are the ways of God, are incomprehensible.

Notes

1. Harry A. Wolfson, *Religious Philosophy*, Harvard University Press, 1961. Particularly the first article entitled "The Platonic God of Revelation and his Latter Day Deniers." p. 247.
2. F.E. Browder. "Is mathematics relevant? And if so, to what?" *University of Chicago Magazine* 67: 11–16, 1975.
3. Wolfson, op.cit., p. 8.
4. Spinoza, *Ethics*. See Wolfson, p. 18.
5. Wolfson, op. cit., p. 20.
6. Wolfson, op. cit., p. 2.
7. Wolfson, op. cit., p. 3.
8. Wolfson, op. cit., p. 6.

Further Readings. See Bibliography

H.A. Wolfson (1976, 1979)

The Computer Thinks: An Interpretation in the Medieval Mode

"If we decide to leave the riddles unanswered, that is a choice; if we waver in our answer, that, too, is a choice: but whatever choice we make, we make it at our peril."—Fitz James Stephen

The Text

"The computer thinks. It carries out operations, which, when people do them, are described as requiring thought. It does its work very well indeed. The computer therefore is smart; it can replace people in jobs that are thinking jobs."

Nonsense, say the opponents. On the contrary, the computer does not think. It is stupid. It does only what it is told to do. No more, no less. It is a super slide–rule. It is a big pencil and a big pad of paper. People think. As they live and grow, they adapt. Their thinking reflects this. People are open–ended. The computer is a narrow mechanism on a fixed track.

This second view is then rebutted: "You are not up to date in your information. The computer *can* change its own program. It can be programmed to learn from experience, to react to a changing environment. It can even have a probabilistic or chance element built into it. Now what do you say about rigidity?"

Then there is the rebuttal to the rebuttal. "There are no programs in existence and none on the drawing board that can exhibit the adaptability of a cat or even of a cockroach. Moreover, if the computer learns and changes, the ways in which it does so are themselves fixed and limited. Even the chance elements are not really chance. They are pseudo-chance and are strictly determined by the formulas that are employed to generate them."

And so it goes. Thrust and parry. Thrust and parry. The computer as a thinking machine is one of the great ideas of today's technological world. Mind inside matter. A decade or so ago some of the philosophical issues were argued at length by the scientific community without, naturally, coming to any conclusion. Isn't it always thus with philosophy? Does the *Theatetus* come to any conclusions about knowledge or does it come only to despair? Today, the thinking computer is something of a *cliché*, a lemon from which all the juice has been pressed. Hardworking scientists and technologists tend to relegate discussions of it to the trashbin of metaphysical irrelevance. If they are consumers of computation, they do not worry about whether the computer, when delivering an answer to them, is thinking. If they are computer scientists, working to expand the potentialities of their instruments, they do as they are able to do, and regard philosophical questions as unproductive.

Does the computer think? Today this question is in the doldrums. Ho hum, most of us would say. Even science fiction is beyond this discussion. Robert A. Heinlein, whose science fiction it is said, "lies at the very center of American beliefs and values," accepts computer thought, but he does not force the issue. In his books, the computer is an entity intermiscible with humans, and the emphasis is on the biological and the mythological. In "Time Enough for Love" (1973), we are told of a computer, Minerva, who has become a flesh and blood woman cloned from twenty–three select parents. Minerva has twinned herself as a replica computer, etc., etc. One of the most popular stories of current science fiction—"Dune"—avoids the question by placing its action in the "post–computer era."

But the question won't go away. It is, in fact, one of the more interesting questions, and it can be related to almost all questions in philosophy and in the philosophy of science. I should like to stir up the discussion a bit. While waiting for more developments in computer science, in brain physiology, in neuro–biology, in the psychology of perception and cognition, developments which may drastically alter previously held positions, while waiting for revised philosophical judgments, I should like to discuss the question in an unconventional manner. I am going to try to adopt the point of view of a medieval commentator and approach the sentence "The Computer Thinks" as such a writer might have approached a sacred text or a line from an epic poem.

"Hold on," you might say, " 'the computer thinks' relates to concepts in contemporary science and technology. It should not be treated in a frivolous manner, nor in a fanciful or archaic manner, in a manner that draws on the tradition of a dying literary–religious humanism." I answer that science is science, but thinking about science and relating science to the universe about us is more a matter of philosophy and metaphysics and poetry. Positivism is expiring, and metaphysics is again, tentatively, timorously, recognized as playing a role in concept formation. Underlying all scientific endeavor is the great metaphysical principle that the universe is understandable.

Medieval Exegesis

In a letter to Can Grande della Scala, Dante Alighieri (1265–1321) writes that he would like the Divine Comedy to be interpreted in two ways: in the literal way and in the symbolic way. All great literature, he says, should be interpreted in these two ways. The mind comes to an understanding of them because they are analogous to the two stages involved in any act of creation, whether of God or of man.

Dante further subdivides the symbolic mode of interpretation into three: the allegorical, the moral, and the anagogic. The anagogic is less familiar to modern readers and requires a word of explanation. It relates to mystical and spiritual uplift. In the anagogic mode, people, events, or institutions are interpreted as foreshadowings of the future state of bliss awaiting the faithful. One medieval commentator summed it up neatly: the literal teaches us what happened, the allegorical teaches us what to believe, the moral teaches us how to behave, and the anagogical teaches us what to hope for.

To illustrate how one statement may be expounded in these four ways, Dante selects verses from Psalm 114:

> "When Israel came forth from Egypt
> And the house of Jacob from a people of strange speech
> Judah became His sanctuary
> Israel, His dominion."

Focussing on the first mode of interpretation, Dante says that the literal meaning is that, simply as a matter of history, in the time of Moses the people of Israel who were enslaved in Egypt were redeemed and were let go. In the allegorical meaning, this lines refers to the salvation and redemption that is worked by Christ. The moral meaning is that the soul can be *converted* and moved from the misery and grief of earthly sin to a state of grace. Finally, in the anagogical meaning, the *holy* soul is now redeemed from the slavery of this corrupt earth and is translated to the liberty of life eternal and glorious. In this way, four interpretations are established corresponding to the four notions of the word "redemption."

As an aside here, it should be pointed out that the formation of categories is not alien to contemporary philosophical thought. Thus, Sir Karl Popper, a leading philosopher of science, also discovers that there are three worlds:

> First, there is the physical world—the universe of physical entities—this I will call "World 1." Second, there is the world of mental states, including states of consciousness and psychological dispositions and unconscious states; this I will call "World 2." But there is also a third world, the world of the contents of thought, and indeed, of the products of the human mind; this I will call "World 3."[1]

The computer thinks. It is clear that one cannot really discuss this sentence using the ingredients that were vital to medieval thought. We have our own ingredients. But I shall retain Dante's format and discuss the statement, taking it in the literal sense, in the metaphorical sense, in the moral sense, and in the anagogical sense. As is traditional, I begin with the literal.

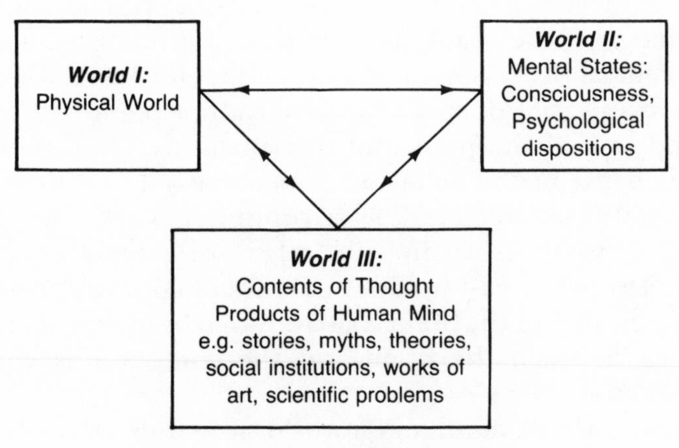

The three worlds of Karl Popper.

The Literal Mode

The computer thinks. Even as you and I think. You and I play chess, the game of reason par excellence. The computer plays chess better than most people. Therefore the computer thinks. You and I do arithmetic. When we sit down with a pencil and paper and do it, we have to think about what we do. We have to remember the addition and multiplication tables. The computer does arithmetic faster, better, perhaps a billion times faster. Ergo, the computer thinks. The air controllers at the airports have jobs that are largely mental. They suffer mental stress, for example. The computer has been programmed to provide landing systems for airports. Ergo, the computer thinks.

The computer thinks in order to overcome the inadequacies of our thought and to replace our thought with a much more efficient kind of thought, to replace human genius with a kind of automated, symbiotic genius.

What is behind this position? *C'est pour épater le bourgeois*, of course. First of all, there are the great successes of the computer in carrying out certain operations that have been associated with mental abilities and there are the prospects of many future such successes. The computer can do arithmetic, and it can carry out numerous processes of higher mathematics as well. Computers were first designed entirely for the purpose of doing mathematical calculations. Today, only a small fraction of computer time is spent in doing mathematics. Simulations of wider varieties of mental activity have become predominant.

Secondly, there are certain superficial global resemblances between the physiology of the brain and the organization of a computer. The brain is a central processor and control unit. Of course, if we split open a computer and split open a brain, we cannot point to a one–to–one correspondence of the parts and the functions. One thinks one sees similarities between the behavior of neurons and that of logic gates. We understand completely how a computer works; we understand perfectly—or we think we do—the relationship between software and hardware. Despite a vast amount of physiological information, however, the operation of the brain is surrounded by question marks; hence the analogy between the brain and the computer is equally questionable.

Is there a world of mind and a world of matter? If so, how do they interact? Or is this categorization and separation a delusion and a trap? The idea that the brain is a computer relates intimately to the view that man *is* a machine. The universe is material, it is mechanism, and man, as part of the universe, is also mechanism. The statement that the brain is a computer, at the moment, refers to a research program, a hope for the future.

There is mind and there is matter. How can the two be unified? Nobel Prize winning biologist George Wald points out that although we know vast amounts about the physiology of vision, we do not know what it means "to see." One of the dreams of biology, he says, is the acquisition of this kind of knowledge.

Examine the vocabulary of the world of mental processes. It contains such words as "thought," "intelligence," "recognition," "perception," "cognition," "awareness," "self–awareness," "reason," "dedication," "memory," "intuition," "idea," "belief," "pattern," "relationship," "urge," "beauty," "explanation." There are hundreds of them. We should like, if possible, to correlate these terms with what goes on physically and chemically in the central nervous system. In addition, we should like to correlate these mentalistic terms with patterns of behavior. We also must correlate, as the third side of the triangle, the physics with the behavior. All three problems are notoriously difficult, and, of the three, the first has perhaps been pursued least successfully.

Now comes the brash new computer science. It has its own panoply of terms: languages, programs, loops, feedback, information, data structures, zippered lists, accessing, databases, protocols, parallel processing. These ideas seem to form a bridge between purely mental conceptions and purely physical. Perhaps they offer a possibility of constructing such a bridge, or even of eliminating the dichotomy between mind and matter. "The computer thinks," interpreted literally, expresses the hope of an ultimate unification.

□□
□□

Turning to Professor Allan Perlis of Yale University, a distinguished computer scientist, we find that he states categorically, "The only difference between living and nonliving matter is in chemistry," and adds that the proof of this statement constitutes one of the major dreams of computer science. Computer science, mind you!

Next let us turn to behavior. Suppose I say, "the computer thinks." What do I really mean? I think you think. I suppose, though you and I are not built in an entirely identical fashion, that what is happening when I think is pretty much the same as what is happening when you say you think.

I suppose this, but I am not entirely sure. A note of poignancy is added to this doubt when one considers mental illness. Is a man crazy or is he only pretending to be? If he keeps pretending to be crazy, then isn't that in itself a form of craziness? From the gap created by this doubt, we may come to see a need for definitions.

The famous Turing test, proposed by Alan Turing, one of the geniuses who founded theoretical computer science, goes something like this.

Let a person *A* interrogate a computer program *C* and interrogate another person *B*. Both *B* and *C* are hidden from *A*. They communicate with *A* by a terminal. Now, if *C* can convince *A* that *B* is a woman, then *C* can be said to be thinking.

Turing showed how this could be achieved. There is no doubt it can. You and I are often fooled by recorded messages over the phone. People have been fooled by a computer system for psychoanalysis; or, if not fooled, at least put into a state of mind conducive to a continuation of the "psychoanalytic" dialogue. A computer can be programmed, can be honed up to perform limited but specific tasks, tasks that we ourselves have formalized. Is this thinking? A Cuisinart can slice carrots, a task not performable by the Headless Horseman, but is the Cuisinart thinking? Was a Chinese fortune cookie thinking when last week it provided me with a piece of good advice?

"Computers think." This is absurd. I say it's absurd. You may say so. Karl Popper says it's absurd:

> I would say without hesitation that they can not, in spite of my unbounded respect for A.M. Turing who thought the opposite. We may perhaps be able to teach a chimpanzee to speak—in a very rudimentary way; and if mankind survives long enough we may even speed up natural selection and breed by artificial selection some species which may compete with us. We may also perhaps be able in time to create an artificial microorganism capable of reproducing itself in a well–prepared environment of enzymes. So much that is incredible has happened that it would be rash to assert this to be impossible. But I predict that we shall not be able to build electronic computers with conscious subjective experience.
>
> Turing said something like this: specify the way in which you believe that a man is superior to a computer and I shall build a computer which refutes your belief. Turing's challenge should not be taken up; for any sufficiently precise specification could be used in principle to programme a computer. Also, the challenge was about behavior—admittedly including verbal behavior—rather than about subjective experience.

"Computers think." Absurd? Perhaps. But there is a group of computer scientists, brilliant, creative, who do not think it absurd. They feel that they are brothers and sisters under the skin to equally brilliant and creative bits of wire and ceramic. If we deny their kinship, are we playing a role similar to that of Bishop Wilberforce, who denied Darwin's claim that man is related to the ape?

Perhaps the problem is only semantic: there are sixty–nine interpretations of the verb "to think" and every single one of them is right. Some of them lead to the statement "the computer thinks," some to its denial. The advocates of the literal view serve us notice that as the computer duplicates more and more types of human mental activity, it gets closer and closer to displaying all sixty–nine interpretations of thought.

The deniers of the literal view say: "computer thought only goes up to a point. Beyond this it cannot go." They point to the lesson of Otos and Ephialtes, who tried to scale Olympus and kill the Gods, and who themselves were slain at Naxos. "The computer has no self–consciousness, no subjective experience, no belief, no originality, no intuition, no love, no telepathic or psychic awareness, no soul." The rebutters of the deniers attend international conferences on artificial intelligence and present papers with such titles as "Why Robots Will Have Emotions," and "Ascribing Mental Qualities to Machines."

Computer Science Bridges the Gap?

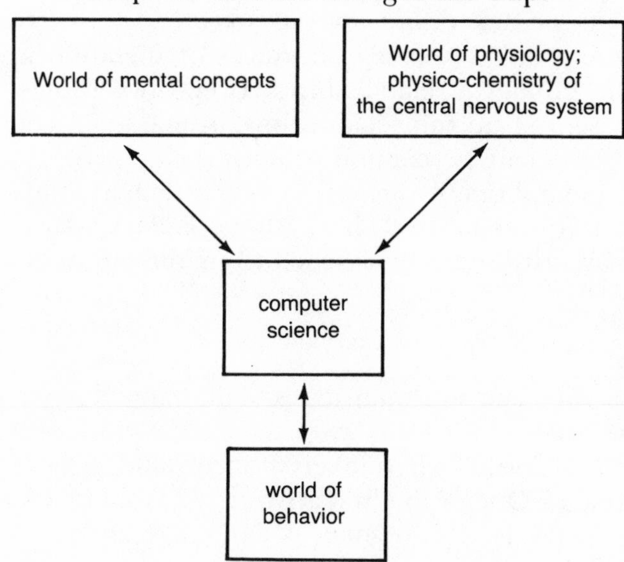

Now go all the way, and ask the question: are computers alive? For, after all, is not life a precondition of thought? The epateurs say: "yes, of course, computers *are* alive. Do they not satisfy the four criteria for life? Do they not (1) exhibit characteristic forms and organizations, (2) process energy, (3) exchange information and (4) reproduce? (As the bees are to flowers, so is man to his computers.)" Q.E.D., assuming that you agree to the reduction of life to these four sentences.

The Metaphorical Mode

The computer thinks. Metaphorically, that is.

Let's begin with some observations on metaphor itself.

> "The morn, in russet mantle clad
> Walks o'er the dew of yon high eastern hill"—Hamlet

The morn does not wear a reddish–brown coat.

> "The trumpet set all the shores ablaze with its sound"
> —Aeschylus, Persae

Trumpets cannot ignite shores.

> "Strip off the garments . . . and reascend in our naked selves"
> —Plotinus, Ennead

Our bodies can be naked, but not our selves.

What is going on? These are instances of metaphorical language, language that implies a relationship between two things, a similarity of a sort. In setting up the relationship, in making a comparison between two things, our perception of both is changed.

There are metaphors of various types. Particularly important is the type called by John Ruskin "The Pathetic Fallacy." In this figure of speech, human attributes are ascribed to inanimate nature:

> "The sea rages."
> "The flame devours."

The sea does not rage as a man rages. The flame does not eat as one who hungers eats.

If metaphor is absurd when taken at face value, why do we use this kind of language? One of the first answers was that given by Aristotle. Metaphor, he says, is an ornament of language, a decoration. It perceives and states analogies, and in this way it intensifies expression. It produces an emotional charge.

"The damp souls of housemaids sprouting despondently at area gates"
—T.S. Eliot

Insofar as poets seek heightened language and intensified feelings and understandings, metaphor is indispensable for them. Insofar as poets want "to utter the unuttered and to name that which has not yet been named," they are driven to metaphor, but metaphor plays a much more fundamental role than this. It is much more than decoration. Metaphor gives us a way of organizing the world. Diverse phenomena are linked without sacrificing the diversity of the components. The world is dynamic. One experiences both being and becoming. Only the fluidity of metaphor enables language to adapt itself to our perception of a changing, multi–dimensional world. Just as in nature new things emerge from the combination of basic elements so in language metaphor creates new meaning by juxtaposition and synthesis. Out of the tension between the literally false and the metaphorically true comes new insight.

Some extremists say that at bottom all language is metaphorical. There is no language that is infinitely pure and infinitely precise. This kind of high fidelity is reserved for the Deity, or for the Deity's alter ego—mathematics.

Some scholars distinguish what they call "root metaphors." These are metaphors that are so forceful and convincing that they can affect the way we live our lives.

The World is a Garden.
The World is a Stage.
"All the world's a stage
And all the men and women merely players"—As You Like It.
The World is a Vale of Tears.
The World is the Antechamber of the World to Come.
The World is a Machine.
The Body is a Machine.

When one gets an eye from an eye bank or is hooked up to a dialysis device, the literal pushes the metaphorical very hard indeed.

The Mind is a Computer.
The Computer is a Mind.
Mathematics is Deity's alter ego.

If an explanation is given along metaphorical lines does the avoidance of the literal represent a cop–out, a weakened vision? Take, for example, God. Now here, surely, we have one of the most fundamental, constantly occurring, constantly controverted ideas in the whole of human experience. What would a literal interpretation consist of? An old man with a beard, located somewhere in the clouds, controlling the universe and, as in Michelangelo's "Creation," passing the vitalistic spark to Adam? We are not primitives, so that God as an animal (literally) or God as a man (literally) are rejected. When Christians assert the literal identity of God and man they are judged heretical. On the other hand, God as something close to mathematics (Plato, Aristotle), God as the philosophical ideality (Kant), God as Cosmic Ether (Ernst Haeckel), do not send us into sympathetic vibrations. We seem to be forced into metaphor. "This (wine) is the Blood of Christ. This (bread) is the Body of Christ." Is the "is" here literal or metaphorical?

We say: God knows, God says, God hears, God thinks, God creates, God remembers, God judges, God watches, God is jealous, God is a father or a king. God is good. These are all anthropomorphisms. In fact, languages — English, Latin, Greek or Hebrew, for example — have only a few words which are uniquely applicable to God. Omnipotent and omniscient come to mind, and of course the tautologuous divine, godly, and so on.

What say the theologians about metaphorical interpretations of God? On the whole, they like them and promote them. St. Thomas, in his Summa Theologica, supports metaphor, quoting Hosea 12:10 as a proof text: "I have multiplied vision and used *similitudes* by the ministry of the prophets."

Aquinas goes on to say that "to put forward anything by means of similitudes" is to use metaphor.

> It is befitting Holy Scripture to put forward divine and spiritual truths by means of comparisons with material things. For God provides for everything according to the capacity of its nature. Now it is natural to man to attain to intellectual truths through sensible things, because all our knowledge originates from sense. Hence in Holy Scripture spiritual truths are fittingly taught under the likeness of material things.

Nonetheless, for Aquinas, the literal has primacy.

Three centuries later, John Donne (d. 1631), is troubled by the distinction and the split between the literal and the metaphorical. This passage is from one of his Devotions:

> My God, my God, Thou art a direct God, may I not say a literal God, a God that wouldest bee understood literally, and according to the plaine sense of all that thou saiest? But thou art also (Lord I intend it to thy glory, and let no prophane misinterpreter abuse it to thy diminution) thou art a figurative, a metaphoricall God too: A God in whose words there is such a height of figures, such voyages, such peregrinations to fetch remote and precious metaphors, such extensions, such spreadings, such Curtaines of Allegories, such third Heavens of Hyperboles, so harmonious elocutions, so retired and so reserved expressions, so commanding perswasions, so perswading commandments, such sinewes even in thy milke, and such things in thy words, as all prophane Authors, seeme of the seed of the Serpent, that creepes, thou art the Dove that flies . . .
>
> Neither art thou thus a figurative, a metaphoricall God in thy word only, but in thy workes too. The stile of thy works, the phrase of thine actions, is metaphoricall. The institution of thy whole worship in the old Law, was a continuall Allegory; types and figures overspread all; and figures flowed into figures, and powred themselves out into farther figures; Circumcision carried a figure of Baptisme, and Baptisme carries a figure of that purity, which we shall have in perfection in the new Jerusalem. Neither didst thou speake and worke in this language, only in the time of thy Prophets but since thou spokest in thy Son, it is so too. How often, how much more often doth thy Sonne call himselfe a way, and a light, and a gate, and a Vine, and bread, than the Sonne of God, or of Man? How much oftener doth he exhibit a Metaphoricall Christ, than a reall a literal? . . .

Thus, while asserting that God is both literal and metaphorical, Donne the poet is able to give us many more instances of a metaphorical God than Donne the theologian gives us of a literal God. He reduces all to metaphor and allegory, both word and deed, both in the Father and in the Son.

Yet with the possible exception of the deepest Christians, those who are able to believe the Central Mystery of their faith, those who are able to assert that the Incarnation ("The Word was made flesh and dwelt among us") is the culminating metaphor of the Sublime Poet who Himself enacts it in reality, most people feel that what is metaphysical is not real, that metaphor avoids a direct confrontation with reality and is hence a dilution of belief.

The writings, thoughts, beliefs and rituals of religion are among the finest expressions of metaphor. Religious statements are poetic. They are non–interpretable in the literal sense. In this sense, they are even absurd. "I believe precisely because it is absurd," said Tertullian. The metaphysical attributes of God, His omniscience, His omnipotence, His self–awareness, His self–sufficiency, His self–love, His constancy, are said to be meaningless in the sense of logic. Religious notions, such as the efficacy of prayer, man's immortal soul, or future bliss in the world to come, are untestable in the sense of scientific method. Do not teach this stuff, writes Sir Francis Crick, discoverer of the molecular structure of DNA. "Since much of this . . . from the point of view of most educated men is nonsense, it seems to me to be particularly distressing that (religion) should be the one compulsory subject of British education" (*Of Molecules and Men*). A softer view had been put forward a century before by the Irish historian W.E.H. Lecky (1838–1903): "The religion of one age is often the poetry of the next . . . Religious ideas die like the sun; their last rays, possessing little heat, are exhausted in creating beauty" (*History of Rationalism*). Is there then, any fire in metaphor?

In contrast to the statements of poetry or of religion, the statements of science have been held to be verifiable in the sense of experimentation and hence to be true in the literal sense. Not so, say contemporary philosophers of science. Statements of science may be adhered to as long as they have not been *falsified* (and, some say, even longer). Potential falsifiability is the hallmark of the scientific statement. Little by little over the past century and a half, under the impact of such developments as non–Euclidean geometry and relativity theory, scientists have been replacing the idea of a theory of *xyz* with the idea of a model of *xyz*. There is a world of difference philosophically and psychologically between a theory and a model. The word "theory" seems to imply an idea that is *true*. Permanently. The word "model" implies an idea that is *expedient*, but evanescent. Francis Bacon, living on the borderline between the medieval and the modern ages of science, perceived the distinction when he criticized the Copernican solar system as one which sacrificed truth to expediency.

In a model of a physical phenomenon, our perceptions of one part of the world are explained by our perceptions of another part of the world. An analogy is drawn. What other kind of explanation can there be? In many instances, a physical phenomenon, residing in World I (in the sense of Popper) is explained by something else in World I. Consider, for example, the explanation of nature given in the religion of ancient Egypt. The sky goddess Nut arches over the bowl of the heavens, her feet and hands firmly planted at opposite horizons. Her breasts hang down from the center. In some representations she is given multiple pairs of breasts. Below, the earth god Geb fertilizes Nut. Here is a model, a metaphor; and out of it we may make certain conclusions. One can find examples of this kind in contemporary physics.

Very often, a World I phenomenon is explained in terms of World II. The planets obey Newton's laws of motion and gravity.

> Law I. Every body continues in its state of rest, or of uniform motion in a straight line, unless it is compelled to change that state by forces impressed upon it.
>
> Law II. The change of motion is proportional to the motive power impressed and is made in the direction of the straight line in which that force is impressed.
>
> Law III. To every action there is always opposed an equal reaction.

Does the moon obey Law I even as Newton obeyed the laws of Parliament? Nonsense. This is metaphorical language. The verbal description in Law I and the differential equations of motion implicit in Law II are both metaphors. The story of Nut and the laws of Newton are both models. They yield different things. Out of the story of Nut, we may derive moral concern for the processes of nature. Out of the differential equations of meteorology we have not, as yet, derived such concern. Do the planets follow Newton? Perhaps it should be turned around: Newton's laws—and Einstein's—had damn well better obey whatever it is the planets are doing, or their laws are out of business. Chop chop. Falsified. But not even this is the full story.

Do contemporary scientific models represent truth or, as in Lecky's line just quoted, only the poetry of a dying world view?

We have anthropomorphised the computer. "The computer thinks, the computer knows, the computer remembers, the computer learns. The computer says. We load it (as we load a horse or a wagon). We instruct it (as would a patient teacher). We command it (as a sergeant commands a private). The computer fouls up. If we put garbage in, the computer gives us garbage out. The computer is to blame." Here human qualities are ascribed to inanimate nature. Here is the pathetic fallacy, the humanization of the machine. Of course, it can work both ways. There is also the mechanization of humans: "He got up a head of steam. He is turned off, turned on. She was thinking so hard, you could hear her wheels go around. I need a change of oil. His wife has him programmed."

"The computer thinks." Taken as a metaphor, what this language does is to link a certain activity of humans (who surely think) with the activity of the computer, and assert that we have found parallels. We observe that this kind of language not only is colorful, but can elicit productive insights. How far can we push the language? Can we really remove the quotes from "thinks"?

Why do we anthropomorphise the computer? Many reasons come to mind. It is a way of expressing and increasing the warmth of the man-machine relationship. We do it with boats ("She's a sweet little thing, and has clean lines"). We do it with weapons (King Arthur's Excalibur, Big Bertha). This language also is a method of dealing with our frustrations. When our check hasn't come through, there is no point in kicking the central processing unit for what is—ultimately—our own program's mistake. Instead we say "the computer has fouled things up"; we grin and bear it.

Other kinds of intensification come from the anthropomorphism. When we create structures, we want to have confidence in them: in our houses, bridges, the family, the state, our religious and philosophical beliefs. We would like them to be a firm basis on which to build a life. In the last three hundred years, change has been rapid, structure after structure has cracked to the ground. Despite protestations to the contrary, despite persistent cynicism, we have a residue of belief in rational thought and in progress by its help. Although the electronic computer is new, its underframe is mathematics, an ancient doctrine that has always symbolized the rational element of the universe. The computer, at the moment anyway, symbolizes progress through rational means. When in doubt, compute.

"The computer thinks." Adopt this stance and we benefit in many ways. Because this statement is metaphorical, does it follow that important physical consequences are thereby precluded? Not at all. As we have seen, all scientific explanation has a metaphorical quality. By arguing from how a computer works, and by utilizing the results of the computer in studying mathematical models, we may conceivably learn things in neurobiology and in psychology. The computer may lead us to knowledge about cognitive processes, about memory, and about whether the hardware/software split is applicable to the human brain. Perhaps—who knows—it will lead us, by analogies, to insight on mental illness.

This is a hope, and it leads us to the moral mode.

The Moral Mode

The computer thinks, and by its thoughts we are elevated. By its good deeds we are bettered.

The utility of the computer, actual or hypothesized, is so much on everyone's mind that the reader will have no difficulty in arriving at a personal judgement of its accomplishments and shortcomings. From Pac Man to pacemakers, from trips to the moon to digital recordings, from election predictions to stylistic analyses of the *Federalist Papers*, from mathematical physics to new discoveries in mathematics itself, the world has moved into an intensified phase of mathematization, and the computer is the engine that pulls and pushes. Its successes have given it a blind, euphoric drive that makes it say: all, all is grist for my bits and chips. The amount of money that has been made is enormous; the amount spent on publicity and public relations has been enormous; the number of words both of sense and of drivel elicited by the computer as a commercial object is enormous. The computer touches everyone.

Have you visited a computer store? It is surely a harbinger of a new era when one encounters the spiritual descendants of snake oil salesmen pushing "peripherals," or when one recognizes friendly retreads from Honest John's Used Car Lot high pressuring software. What is good must sell, and what sells must be good. This is the morality of the market.

The present is bright and splendid. And the future? Well, according to the received wisdom, "we ain't seen nuthin' yet." The present state is a drab antechamber to the glorious world of the future, wherein will be found goodies galore and miracles beyond conception. We are promised "intelligence built into most manufactured objects." We are promised "a national information network that any intelligent home will be able to access." We shall be educated to the point where those humans "unaided by computers will appear feeble-minded." We shall be protected by constant vigilance of computers which "by taking adequate safeguards, could prevent us from making dangerous mistakes." These quotations are from the article "The Case For Computer Literacy" (*Daedalus*; Spring, 1983) by John G. Kemeny, who is a mathematician, philosopher, educator, former President of Dartmouth College, member of national commissions, and a man who has been in the vanguard of computerization for thirty years. The admission price

Kemeny wants us to pay for entrance into this state of mental amplification is universal "computer literacy."

It would be an appealing exercise to tear into Kemeny's article (as into any article on computer futureology) and criticize its fatuousness, its lack of documentation, its over–valuation of the present accomplishments, its neglect of the instability brought on by computerization, its naïve faith. It would be easy to criticize the assumption that people must learn to program and learn to enjoy programming and that this activity must replace their beloved carousing and gossiping. It would be easy to wag a finger and say, "you really haven't answered the question 'what will humans do?' because your philosophy is that of expiring positivism, and your vision is limited to technology," but this would divert me from my task which is, in this section, to interpret in the moral mode the statement, "The computer thinks." Surely intelligence, education, information, vigilance are components, necessary but insufficient, of the moral life. Insofar as the computer fosters these things, it is a force for good. What is vital here is to decide in what way the computer—in its *thinking* capacity—is a force for good. An explanation may be attempted along the following lines.

Our own (human) thought, though splendid enough to have removed us from the trees where we sported with our sibling apes, is fragile, inadequate, limited; it is corruptible and often stands naked in its corruption. When we say that the computer thinks, and say it in the moral mode, what can we mean except that the computer, through its unique qualities of rapidity and logical sharpness, brings to the processes of deduction, cognition and creation, ingredients that would elevate these functions. Through our association with the computer we may be enabled to live a more moral life than was hitherto possible.

One may dream of man–machine symbiosis along the lines of Hamlet: "What a piece of work is this man–machine! How noble in reason! How infinite in faculties! In form and moving, how express and admirable! In action, how like an angel! In apprehension, how like a god!" This is the dream. It is the vision of Leibnitz who dreamed of reducing ethics and morality and law to computation, so that when litigants stand before the judge with their causes, the judge will say, "Let us compute," and thereby arrive at a just decision.

And yet, next to the dream stands the reality, and we must examine the record of this quintessence of chips and dust. The computer is programmed to realize certain particular mathematical models of social and economic structures. What answer you get depends upon what model you run. What advice you get depends upon what criterion you use of what is good. The computer is asked to simplify our lives; by the computer version of Parkinson's law, the degree of mathematization rises to meet the capacities of the machines available. The computer is an adjunct of the military. Its computer–controlled "smart" bombs are both stable and unstable in their deadliness. Several new brands of criminal activity, practiced by skilled programmers and computer crashers, have been born lately.

The masters of science fiction have written decalogues for the proper behavior of robots toward humans. In comparison to them the simple code of Hammurabi (18th century B.C.) sounds like the complex chaos of today's legal system. If the computer is self–programmable, and if it can learn from the experience of earlier generations, will it not pick up all the nasty little tricks of its progenitors? Or will it, instead, miraculously find the six steps to moral purification which have been conjectured by Lawrence Kohlberg[2] as being innate in *all* humans?

The computer thinks, and we are enriched and ennobled by this thought.

The Anagogic Mode

A small fraction of current research in computer science is in the specialty known as "The Physical Limits of Computation." The practitioners of this subfield, brilliant people with a strong bent towards Cloud-9 speculations, worry about such questions as: if one requires a switching time of 10^{-10} seconds, how large can the computer be? Answer: 3 centimeters. What is the fastest rate at which information (in bits per second) can be transmitted, given the energy available for the transmitting? How much memory can a computer have? Ultimately: how much computation can *ever* be done?

Such questions as these inevitably lead the speculators to relativistic quantum physics, and to the theories of information and thermodynamics. Entering into their rule of thumb calculations are the age of the physical universe (2×10^{10} years), the total mass of the universe (10^{55} grams), and such things as the largest number whose digit string can be stored in a human brain. The calculations are mind blowing: there are some mathematical problems that are so complex that using the best available methods and converting the whole universe from its inception into a computer to do the job, an answer would still not be forthcoming. Such problems are known as transcomputable. They lead to the basic question: should a physical theory be a theory that is computable? Which of them are?

These speculations are fascinating, and paradoxical, but they cannot be pursued here. I have exhibited them for another purpose. The physics leads us directly from concepts of day to day proportions, to speculation about the atomic and the galactic, to the microcosm within the microcosm. Ultimately, if one thinks long enough in such a vein, one comes to the position that all may be computation, and that the whole universe is doing mathematics.

This is not a new doctrine. It is at least as old as Pythagoras (550 B.C.). By the time we arrive at Pseudo–Aristotle (*De Mundo*: c. 50), cosmology runs along these lines: God is the unmoved mover, fixed, immutable. God is mind, or thought. Thought, in its purest, most refined version is mathematics. All things have a rational cause, which is to be found in God. God thinks, and the world, which is a huge machine, is driven by His thought. His thought embodies the principle of the reconciliation of opposites, male and female, day and night, dry and wet, heavy and light, hot and cold. These qualities are allotted and balanced against one another in precise mathematical form.

At the first level, the motions of the planets, of the sun and the moon, are affected. Read these motions properly and one reads God's mind. The future stands revealed. In what way do we read the motions properly? To this question there are at least two answers: the way of astrology (which was a serious option until about 1600) and the way that evolved into what has become normal contemporary science.

To think, in the divine sense, is to compute, and to perform earthly computation is to plug in to the divine plans. The statement that the computer thinks can be interpreted allegorically as saying that God's Providence governs the world mathematically.

I doubt very much that any of the computer scientists who assert that the computer thinks have, in their private devotions, turned data processing and number crunching into moral theology. I pass over this interpretation lightly and proceed to a final interpretation where the assertions are more explicit and rather more anagogical.

In 1963, Professor I. J. Good, currently at Virginia Polytechnic Institute, wrote a remarkable paper entitled "Speculations concerning the First Ultra Intelligent Machine." Put out under the imprint of the Institute for Defense Analysis at Princeton, the Admiralty Research Laboratory at Teddington, England, Trinity College, Cambridge, and the Atlas Computing Laboratory, Chilton, this paper gives us simultaneously intimations of two states: a State of Ultra Intelligence and a State of Salvation.

Professor Good's scientific credentials are impeccable. A brilliant mathematician and statistician with hundreds of important publications, a philosopher of science, in his younger days he was a member of the famous Bletchley group of cryptanalysts. Good observes first of all that there is no point in building a machine with the intelligence of a man, "since it is easier to construct human brains by the usual method." One should therefore build an ultra–intelligent machine which may "be defined as a machine that can far surpass all the intellectual activities of any man however clever." Now, this paper does not constitute a blueprint for such a machine—naturally. It is only a long series of speculations on what the possible ingredients for such a machine might be and what difficulties stand in the way of its construction.

The jump which enables us to transcend human intelligence remains fuzzy. It must be so, for to explain it completely would itself be an act of ultra–intelligence. Well, almost. One hopes that the jump will come about by the juxtaposition of vast quantities of good hardware and an appropriate amount of good software. Mix well, and hope that the mass will go critical and constitute ultra–intelligence.

The first ultra–intelligent machine, says Good, will be the *last* invention that man need ever make, for the machine will take over, providing us with the super–ultra intelligent machine. In this way, we will have set off an intelligence explosion whose consequences are indescribable.

Now, Good is not the first person to have imagined such a thing. Perhaps he is one of the first to accompany his speculations with so much probability theory. His article is conveniently at hand so I used it as a prototype.

Shall we classify it as science or science fiction? Why bother to draw the line? For as Good says, "It is sometimes worthwhile to take science fiction seriously." Whatever it is, the reader comes to believe that Good is laying this out almost in the same spirit as the famous argument given by Pascal for believing dogmatic theology: make a wager that God exists. If he doesn't, you haven't lost a thing, but if he does, then your belief will entitle you to the rewards of infinite bliss in the world to come.

If ultra–intelligence is not an obtainable goal, Good seems to be saying, then I haven't lost much. If it is obtainable, then I will have been one of the first to point it out—even if my specifics are inchoate. Perhaps he had in mind the famous letter from Einstein to Roosevelt warning the President of the high probability that an atomic explosion was possible. Perhaps Good had in mind Descartes' famous universal method. This was intended to unify all scientific discourse and research and was to be as far beyond current method as "are the orations of Cicero beyond the babblings of an infant." Now the ingredients of the method as reported in Descartes' *Discours de la méthode* are vague—except in geometry. There they have been so fruitful, extending to all of mathematics and to science in general, that it would not be a mistake to call our age and all its scientific aspirations Cartesian.

Now, imagine the ultra–intelligent machine in place and running. What will it do for us? Does Good tell us? Yes and no. There is little point in saying that it will lead us to the solution of this or that problem, say, in medicine, where there is a reasonable probability that we can make a break–through using our own brains and occasional help from a computer of conventional power. Nor does it make sense to say that it is the last invention that mankind ever need make, for to exist is to struggle, to change, to grow, to adapt, to invent. Our root metaphor now may well be that of Samson Agonistes: Eyeless in Gaza. Shall it be replaced with Mindless in Paradise? The ultra–intelligent machine, in this sense, snuffs out our existence as a stiff wind a candle.

The ultra–intelligent machine must tackle problems that are so difficult that we have lost all hope of our solving them without it. The ultra–intelligent machine must confront Hell on Earth and wrestle with it. This is stated explicitly on the first page of Good's article: "The survival of man depends upon the early construction of an ultra-intelligent machine."

Man has been brought to such a predicament that the only hope for his survival is for him to create a transcendental instrumentality.

The computer thinks. "I know that my redeemer liveth," said Job, to which the New Job added: His name is King Messiah Ultra Computer. It supercomputes and will transfigure mankind into bits.[3]

It ultrathinks and *we* shall be saved. We *shall* be saved.

O sancta simplicitas!

Notes

1. K.R. Popper and J.C. Eccles, *The Self and its Brain*, p. 38.
2. *The Philosophy of Moral Development*, Harper & Row, 1981.

3. "That such a profound human transfiguration might occur by the close of this century using a technological base only relatively little advanced over the existing one is most striking. I therefore suggest to you in conclusion that only the near-term aspects of dealing with the information onslaught are of human interest, and that the next step is about to be taken in the eons-long procession of species: we will as a race be so altered so soon by coping with the information onslaught that taxonomists a century hence will declare *Homo sapiens sapiens* to be extinct. For the first time in the history of life on this planet, though, the extinction of a species may be voluntary, moreover on an individual-by-individual basis, and within a very few generations. Neither artificial nor natural intelligence may be found on Earth a century hence, though intelligence of degrees of which we do not presently dream may then grace the home planet of Man."—Lowell L. Wood. In D.M. Kerr et al., p. 155.

Further Readings. See Bibliography

S. Lem (1976, 1979); J. Eccles and D. Robinson; K. Popper and J. Eccles; A. Allen; C. Borst; H. Dreyfus; R. Gregory; S. Jaki; E. Kent; P. McCorduck (1979); M. Minsky (1975); K. Popper and J. Eccles; B. Raphael; M. Ringle; R. Schank and K. Colby; J. Searle (1981, 1984); G. Simons, A. Sloman and M. Croucher; M. Arbib, D. Berggren; J. Carbonell; F. Crick; D. Dennett; G. Lakoff and M. Johnson; E. MacCormack; J. Kemeny; L. Kohlberg; (Pseudo-) Aristotle; S. Brams; I. Good (1964, 1965, 1977); D. Kerr; R. Landauer (1967, 1984)

Mathematics and the End of the World

IN MY OPINION, in this year 1986 (common era), 5746 years since Jehovah's alleged creation of the universe, it is impossible to write or think seriously on any topic at all without having in mind the highly possible rapidly approaching end of life on Earth.

We need not here rehearse the all too well known details about multiple overkill, nuclear winter, mutually assured destruction, balance of terror, etc., or the forthcoming laser–equipped nuclear–armed satellites of the promised star wars era.

We take all this technical stuff for granted.

Neither will we go into the political side. Our readers, we presume, are already familiar with proliferation, superpower negotiations and breakdown of negotiations, proposals for nuclear freeze (denied) and proposals for nuclear rearmament (adopted).

We take all that for granted too.

All we can talk about here, so tangential as to be almost absurd, is our special little angle, mathematics and the end of the world.

From the viewpoint of the end of the world, the fact that mathematics had something to do with it is bound to seem petty if not ridiculous, but from the viewpoint of mathematics and mathematicians, it is far from trivial to be aware that we made an essential contribution to the end of life on earth.

Of course, it is the physicists who usually accept the greatest piece of guilt, but everyone knows the physicists couldn't do what they do without mathematics, both the "classical" mathematics of prior generations and the "modern" mathematics of Ulam and von Neumann.

Nuclear weapons, from the *Fat Man* and *Little Boy* that devastated Hiroshima and Nagasaki up to the MX and Cruise of our own day, are mathematical objects. They are saturated with mathematics through and through. Painful though it is to say it, they could be regarded as the culmination of a great part of the mathematics of our time and times past.

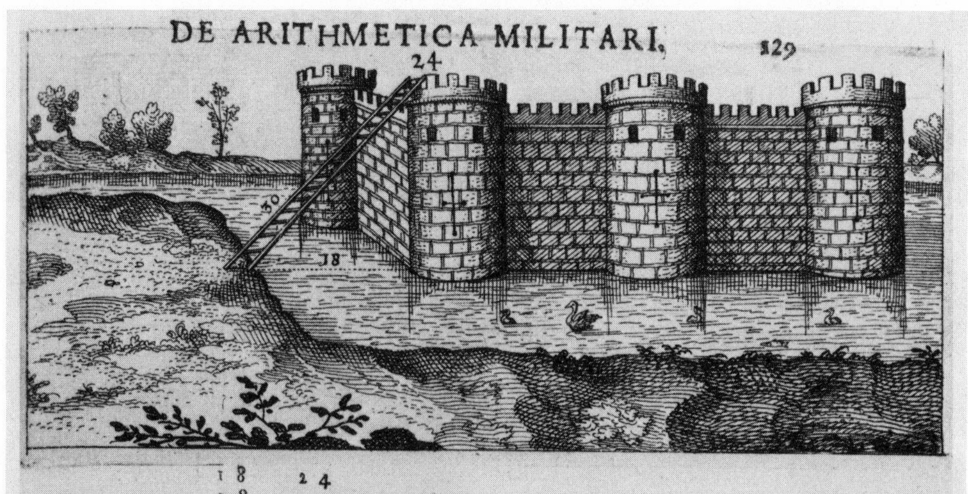


```
 18     24
 18     24     324
144     96     576
18     48      900  (30
324    576     9
```

Regula XIII.

Omne latus superficiei trigonalis, tetragonalis, pentagonalis & consimilis in 6. aequales divisum partem sextam tribuit vice versa numeratam, hoc est, partes ejus extremas ad punctorum calcarium sive humerorum propugnaculi structuram dabit, hoc modo.

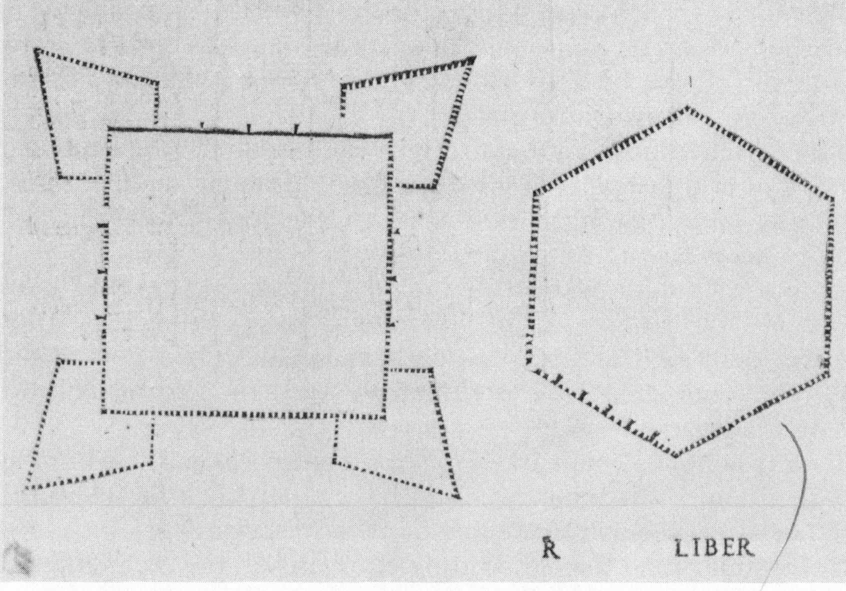

Ducks and Pythagoras' Theorem both decorate a militarized landscape. (*Courtesy of the Lownes Collection, John Hay Library, Brown University.*)

It is true that only a relatively small proportion of mathematicians have participated directly in developing weapons of mass destruction. A giant weapons lab like Los Alamos, with thousands of engineers and physicists (I think), has only a few dozen mathematicians (I believe). (Obviously, this information is "classified" and not available to me except by guesswork.) Nevertheless, Los Alamos could not exist without the complicity of the American mathematical profession. For one thing, all those engineers and physicists had to receive mathematical training in order to help build, update, remodel, renovate and improve the Bomb. (Of course, we have our opposite numbers in Novosibirsk—excepting the great Andrei Sakharov, who finally refused, even perhaps at risk of his life.)

This painful reality was unexpectedly brought home to me recently in the course of my duties as an instructor of undergraduate mathematics.

In February, 1985, a new institution appeared on the mathematical scene of the U.S. It is an undergraduate competition in mathematical modeling, sponsored by "COMAP," the Consortium for Mathematics and Its Applications.

There are other undergraduate math competitions. The most famous is the Putnam Competition, now nearly fifty years old. It follows a familiar format. Each contestant confronts five difficult problems in pure mathematics. The contestants work for two sessions of three hours each, under the eye of a proctor. The premium is on knowledge, ingenuity, self–reliance, and speed.

The new competition is quite different. The students work in teams of three. Rather than trying to solve five hard problems in a few hours, they choose a single problem, and work on that for four days.

The problems among which they choose are "open–ended." A phenomenon or a situation is presented that could occur in real life—for example, in city planning, or in physiology, or in engineering. How would you describe this situation mathematically? Then how would you use your mathematical description to study or control the phenomenon or situation?

My department chairman asked me to organize our participation in this competition. I was reluctant, for I have other work to do. Still, I agreed, because it seemed to me that in many respects the new contest is much healthier than the old. Putting emphasis on the mathematical description of a real–world problem—mathematical model–building

—is a badly needed corrective to the purist slant of the traditional math major. Teamwork, taking as much time as necessary, using whatever tools or resources are appropriate—all this is healthy and good.

The contest did not offer any prizes. I wasn't sure how many of our math majors would want to spend a weekend and a half on mathematical modeling. It was a pleasant surprise to see half a dozen of them—enough for two teams—waiting in my office on the morning of Friday the 15th. There we performed the grand ceremonial opening of the envelopes, which contained the two problems, *A* and *B*, between which each team would have to choose.

Problem *A* was brief. It asked the contestants to choose some species of wild animal, whether bird, fish, or mammal, and to develop a theory of optimal harvesting for that animal.

In other words, develop by mathematical reasoning a policy of how many should be caught or killed each year in order to maximize some appropriate social benefit.

Problem *B* was entirely different.

It covered three pages. I quote here only page 1:

> "Problem *B*. Managing a Strategic Reserve.
>
> "Cobalt, which is not produced in the U.S., is essential to a number of industries. (Defense accounted for 17% of the cobalt production in 1979.) Most cobalt comes from central Africa, a politically unstable region. The Strategic and Critical Materials Stockpiling Act of 1946 requires a cobalt reserve that will carry the U.S. through a three-year war. The government built up a cobalt stockpile in the 1950s, sold most of it off in the early 1970s, and then decided to build it up in the late 1970s, with a stockpile goal of 85.4 million pounds. About half of this stockpile had been acquired by 1982.
>
> "Build a mathematical model for managing a stockpile of the strategic metal cobalt. You will need to consider such questions as: How big should the stockpile be? At what rate should it be acquired? What is a reasonable price to pay for the metal? You will also want to consider such questions as: At what point should the stockpile be drawn down? At what rate should it be drawn down? What is a reasonable price to sell the metal? How should it be allocated?
>
> "The accompanying sheet has more information on the sources, cost, demand and recycling aspects of cobalt."

The other two pages gave some data on the cobalt market, in the form of three graphs and some additional figures.

Each of my two teams had to choose one problem to work on. I was distressed to find out, three days later when they turned in their work, that both had chosen the cobalt problem.

I did not quiz any of the students as to why they made that choice. I suspect that it seemed easier to get started on the problem where they were given the most information at the outset. I'm sure it's not true that they are more interested in military problems than in environmental or ecological ones.

Regardless of how it seemed to my contestants, I, their faculty adviser, felt angry. I had volunteered my time and effort in the assumption that this contest was in accordance with my beliefs and ideals as a teacher, and highest among these is the belief that our mathematical training should serve to benefit mankind, certainly not to destroy it.

As far as cobalt is concerned, I gladly admit that I do not know why the government wishes to stockpile it in case of war. I do remember that some 15 or 20 years ago, some genius produced a calculation to show that radioactive cobalt had a particularly pernicious and deadly radiation, with a half–life just right for maximum slaughter: not so short that it would disappear while people were still alive; and not so long that the radiation would be wasted after the people had died of natural causes.

Consequently, the proposal was made then to produce a "Doomsday Weapon," a giant H–bomb saturated with cobalt. It doesn't have to be shot at the "enemy." Just blow it up on the ground. The radioactive cobalt will drift off all over the world soon enough, and wipe out the whole human race.

The idea of a Doomsday Weapon was the theme of the famous movie, "Dr. Strangelove," some twenty or so years ago.

This contest problem comes 20 years later. I don't know if the government's stockpile is for the manufacture of Doomsday Weapons or for something else. It doesn't matter. The question is, how could such a perverted, anti–educational problem be included on a student contest?

I didn't want to look like a crank or a crackpot. Still, I called the director of the contest, and said to him, "I object."

He said, "I feel the same way."

He explained that the cobalt war problem came about as a result of a last–minute rush, almost as an accident. The mathematician who had promised to contribute problem B was late in turning it in. When he finally submitted his problem, it turned out to be the military cobalt

A Topsy Quilt: local order and global chaos. (*Courtesy of Fred Bisshopp.*)

problem. At that point it was too late to turn it down and try to find another one.

My protest was not rejected; it was welcomed, even encouraged! Then why am I not satisfied?

What this little incident reveals is that nuclear weapons mathematics is an accepted component of American mathematical life. It usually isn't even considered controversial. It's just there, alongside of other kinds of applied mathematics, whether they be oil–field mathematics or transmission–line mathematics. That is why it can slip unintentionally and uninvited into an undergraduate competition. It's not hard to guess what message this Problem B conveyed to the contestants. Namely, nuclear weapons research is considered OK by your teachers and their colleagues. The experience of working on this cobalt contest problem will lessen these students' inhibitions about working on nuclear weapons if and when they become professional mathematicians. Therefore, the inclusion of such a problem can be regarded, in effect if not in intention, as a collusion in the nuclear arms race.

It will be objected, what is wrong with teaching nuclear weapons mathematics? If the military want such mathematics, isn't it the job of mathematicians to supply it and to train those students to perform it who choose to do so?

This attitude treats the military as simply one more customer, no better or worse than any other. It ignores a simple fact: the indefinite continuation of the nuclear arms race renders inevitable the destruction of civilization.

This fact is very unpleasant. It is almost impossible to understand it, take it seriously, and act upon it for more than a few days or weeks without returning to primitive psychological defense mechanisms such as denial.

Nevertheless, once this truth is recalled, it is evident that participation in the nuclear arms race is suicide. Encouraging students to do so is more harmful to society than encouraging them to enter the cocaine or white slave businesses, supposing either of these had need for mathematical consultation.

What one can do beyond refusing to participate in nuclear weapons work is a hard question. The political problem of reining in the nuclear arms race involves obstacles of the highest difficulty in both domestic and international politics. At a minimum scientists can insist on telling the truth, both by words and by example, to a public far from eager to hear it: our present course, if unchanged, will destroy us.

Let us do what we can, by example and by public warning, to resist, slow down, or inhibit the rush to Doomsday.

VI

PERSONAL MEANINGS

Stereographic projection of a portion of the hypersphere S^3 into R^3. (*Courtesy of Husseyn Kocak and David Laidlow.*)

Mathematics and Imposed Reality

WHAT IS THERE and how do we know it? What is real and what is the basis of that reality? These simple and perennial questions constitute one of the fundamental problems of philosophy.

What is real? Well, for heaven's sake, all I have to do is open my eyes and ears and report what I see and hear. That is what is real. I see a typewriter and a telephone on my desk and they are real. Through the window I see trees and birds and they are real. Beyond the trees I see the sky and the sky is real. And the sky seems to be just space and space is surely real—isn't it? Hmmmm. If space is emptiness, how can I see it? If it is nothingness, how can I identify it?

I go outside and kick a large stone (imitating Samuel Johnson's wordless refutation of Berkeley's idealism). The stone is real and the resulting pain is real. Back inside, I look out the window and see the sunshine streaming in. An astronomer would tell me that the sun is really not where it seems to be. Because of the finite speed of light, where it seems to be is where it was eight minutes ago. I follow the position of the sun for several hours. It seems to me it is going round the earth. But no, the astronomer tells me the earth is going around the sun. The evidence of the senses may be insufficient to establish reality.

A lady sits on a throne with a crown on her head. Is she a queen or is she merely an actress playing Gertrude, Queen of Denmark? There are appearances and there is reality, and the philosopher Parmenides taught us that we must distinguish between the two. An oar is in the water and it looks broken. Is it really broken or is it merely the refracted image we are seeing? St. Augustine, who discussed the oar, taught us that as the broken image is surely real, appearances may also be part of reality. What the senses register is merely a projection of reality. To achieve full measure of reality, we must in some way, by intellect, by reason, by intuition, or even by computation, transcend our personal experiences.

Computer generated image "1984" by Thomas Porter based on research by Robert Cook. © 1984 Lucasfilm Ltd.; photograph of hand by Gordon Gahan/PRISM. (*Courtesy of "Science '84," published by the American Association for the Advancement of Science.*)

Did the historical events cited in the Apostles' Creed really happen? Is there such a thing, really, as the soul of a person or of a nation? To what does the ego or the id correspond? What is the physical locus of patriotism or of popularity? What is the status of the ether, of phlogiston, of the pneuma, of the elan vital or of the hundreds of other concepts which science in its historic growth has seen fit to discard?

For the visionary poet and artist William Blake, "nothing is real beyond imaginative patterns men make of reality." For Sartre, one must go to the emotions to find reality. For the British philosopher F.H. Bradley, space, time, motion, and change are all unreal, since they all involve us in a morass of contradiction. For the disillusioned or for the downtrodden, the whole of human experience may be unreal because of its enormity. "It is easy to believe that the world to come is a world. What is much harder to believe is that this world is a world." Reality, then, is a complicated notion containing many levels. It defies satisfactory dissection or description.

Most people do not deny reality. Life is real and not a dream, and the objects that surround them, their thoughts and their feelings are all real. There is objective reality and there is subjective reality. What is publicly and universally agreed upon constitutes objective reality, and the very possibility of achieving universal agreement leads to the conclusion that there exists an objective world "out there," independent of human existence or observation. Subjective reality is what is private. Science is concerned largely with general statements about objective reality. In contrast, literature and art are largely concerned with private worlds and the way they interact.

Most scientists do not ponder over reality—that is the domain of the philosophers. Scientists experiment and theorize, working in a scientific milieu they found in place when they were students, and they get on with their jobs as they can. Out of the scientific experience, however, has come a widely accepted view of scientific reality. A fine statement of what constitutes this reality can be found in the marvelously suggestive book by C.C. Gillispie entitled *The Edge of Objectivity*. Scientific reality is that which has evolved to the point where it is the currently accepted basis of application and of further research and speculation. It is that which is generally agreed upon as the best there is. Objective reality has been achieved through fundamental breakthroughs. Thus, the mathematical theories of Galileo and Newton proved to be the gateway to physical reality, the laws of evolution the gateway to biological reality, etc. This position, which seems completely self–evident, should be amplified by taking cognizance not only of the tremendous changes that have occurred in what the accepted individual facts of science are, but of the revolutionary change that occurred at the time of Galileo and Newton as to the nature of scientific explanation itself.

Consider planetary motion, for example. The sun, the moon, the planets, the stars move from hour to hour and from day to day. How do they move and why do they move in the way they do? Ancient astronomers were wonderfully successful in describing how. They succeeded in separating the daily movements from the monthly, the seasonal, the longer term movements and in organizing them into arithmetical schemes which could be used as a basis for prediction.

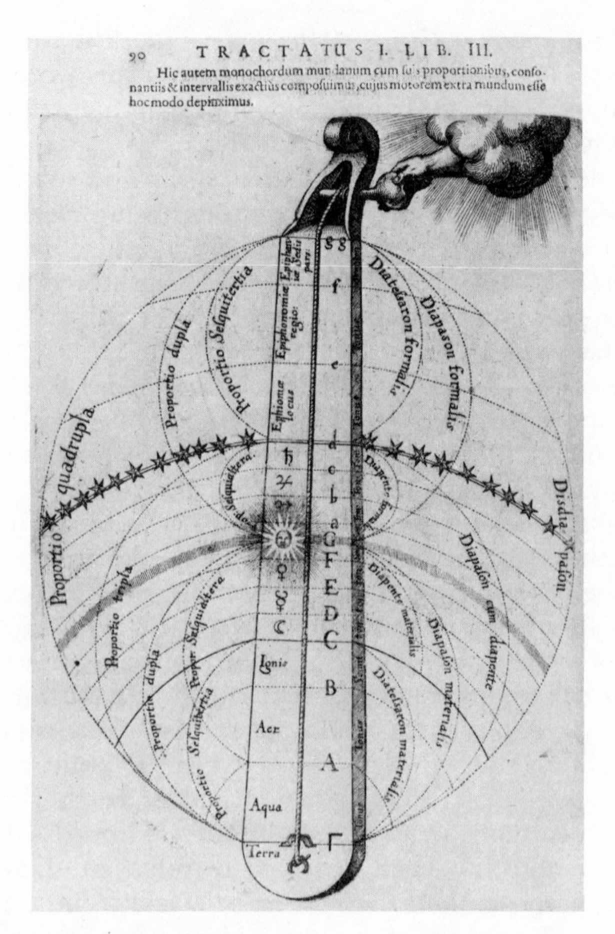

Harmonies of the Universe? A failed conception. (*Courtesy of the Lownes Collection, John Hay Library, Brown University.*)

Our common calendar is a successful arithmetic scheme of this sort. Now, *why* were the movements taking place? The ancient world answered this question in the human terms of goals, values, will, purpose, intent, indwelling tendencies, harmonies and perfections. For the follower of Aristotle, these were the physical realities.

Since the 17th Century, indwelling tendencies and all the others have been discarded. The question of *why* is hardly touched by physical scientists, all the emphasis being placed on *how*; if it is touched upon briefly, as in cosmology, the answers tend to be given in terms of the indwelling properties of the constructs of pure geometry. The metaphysics of the cosmos is given in terms of abstract mathematics which is claimed to be absolutely devoid of goals or purposes: the reality of contemporary cosmology is a mathematical reality.

Gillispie's assertion, that the commonly agreed upon "best of the present" is what constitutes scientific reality, is eminently sensible and agreeable. It allows for history, for change; it looks forward optimistically to the possibility of change, but it lacks one ingredient I believe is crucial. Scientific belief and practice has a component which is *imposed.*

Consider two instances of this. The original edition of Copernicus' "De Revolutionibus Orbium Coelestium" (1543) carried a disclaimer —written by an editor—to the effect that Copernicus regarded his system of planetary movement as being only mathematically convenient but not true in a fundamental sense. One usually tells this story and adds a word about the shameful behavior of the timorous editor who misrepresented Copernicus' intent.

Four hundred years later, famed astrophysicist Sir Arthur Eddington discussed the conservation laws of physics, overtly recognizing the act of imposition:

> "There is no law of government in the external world tending to preserve unchanged specially created entities which occupy it; but the mind has by diligent search picked out the possible constructs which have this permanence in virtue of this mode of construction, and by giving value to these and by neglecting the rest has *imposed* a law of conservation of the things of value."—"The Domain of Physical Science," In: *Science, Religion and Reality,* Joseph Needham, ed. Reprinted, G. Braziller, 1955, p. 217. My italics.

The various laws of conservation are expressed in mathematical language, and their invocation in the practice of theoretical physics is a mathematical imposition which creates a scientific reality. Such laws are often expressed as *impotence principles*: one cannot exceed the velocity of light, one cannot detect uniform motion, one cannot build a perpetual motion machine, etc., and the knowledge of these principles sets limits in our minds as to what can occur in the universe or what we can compel the universe to give us.

It is a clear day and I am flying over the United States. Since childhood, I have carried in my mind a map of the 48 mainland states: a large land mass subdivided into a fixed geometric pattern. I look out of the plane window. Where is the pattern? I don't see it. That pattern

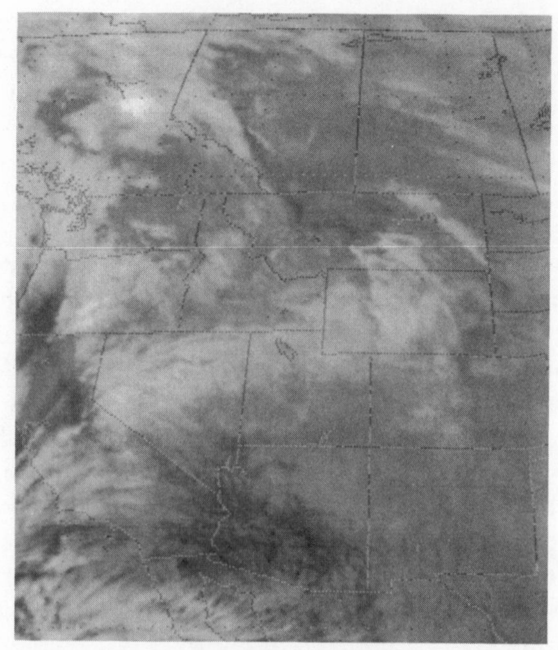

(*Courtesy of the National Oceanic and Atmospheric Administration.*)

is part of our imposed mathematical reality. It has a historical and geographical basis, it has political and economic consequences, but it is largely arbitrary.

There is a widespread sentiment that the way to arrive at objectivity in the real world is to travel the mathematical road. If a subject can be mathematized, this automatically guarantees objectivity. We may, if we so desire, mathematize a good deal, we may introduce measures or criteria, of beauty, of relevance, of connection, of intent, of justice and equity, of economy, of prudence. Even if these measures are arbitrary and inadequate, we can, if we so desire, build them into social and economic structures. If we do this, we create and impose a reality from which we may reap benefits or suffer consequences, as the case may be. Such mathematizations may achieve established status in custom and in law, but they still constitute a reality imposed on human affairs. It is a reality which does not follow logically, as Eddington pointed out, from the "government of the external world." It is not unique: one may imagine alternative or rival mathematizations, equally efficacious, and equally fraught with consequences for the human race.

In days gone by the ideas of intent, purpose, harmonies, imposed a reality on science that was derived from human values. Now, in the reverse direction, science, in its abstract mathematical formulations, has imposed its own reality on human values and behavior.

I conclude with a quotation from Susanne K. Langer. It is a pity that she was not able to write more extensively on this theme.

> "It is hard to realize how long a non–linguistic talent [i.e., mathematics] which is really not at all rare can lie fallow in human beings, only to spring into meteoric career in a few centuries when the right forms of expression are found. . . . We feel that power today as an overwhelming force . . . So great a stride in the evolution of man cannot fail to throw his whole ambiance, social and physical, into convulsion and cause world–wide waves of emotional conflict to build up in every society, savage, barbaric, or civilized . . . It will surely take long and different ages to retrieve the moral and mental balance mankind itself has blasted in the last three or four centuries . . . , and there is no way of guessing whether or how we shall retrieve it . . ."—Susanne K. Langer, Mind: An Essay on Human Feeling, V.3, pp.218–9.

Further Readings. See Bibliography

J.L. Borges; P.J. Davis and D. Park; B. d'Espagnat; C. Gillispie; S.K. Langer; J. Yolton

Loss of Meaning through Intellectual Processes: Mathematical Abstraction

Loss of Meaning

The world is the sepulchre, not only of famous men, but also of dead thoughts, issues, beliefs, customs, rituals, actions. The belief in infant damnation or in the existence of phlogiston must now be explained by antiquaries. The relevance of the Credit Mobilier scandal of a hundred years ago is confined largely to the descendants of Oakes Ames. Professors who march around in academic regalia can now do so only with a good deal of self–consciousness. The divorce of the President of the United States contains no political dynamite. Animal sacrifice is no longer carried out in advanced countries; nor are cuspidors found in well–furnished living rooms, as they were in my childhood. The Divine Right of Kings died under the guillotine with Louis of France. The center of our planetary system has been moved from the earth to the sun. Belief systems are pulverized, and meaning drains from act and object, even as the leaf falls from the tree and turns to dust before our eyes.

Do you believe in witches? Do you believe that it is the duty of a husband to provide for his wife? That the transmigrated souls of humans may reside in cats? Do you believe that it is socially degrading to be in trade? That medicinal cupping is beneficial?

Do you believe that there is a secret meaning to Shakespeare? That the waters of the Black Sea should be whipped when sailors are drowned in it? Do you believe that there is a world to come, better than this world; or that there are a plurality of worlds?

Should one say prayers on the new moon? Should one build a house with a laundry chute? Should one bait bears for sport or write masques in rhymed French verse? Should one teach Latin to young children or serve calves–foot jelly for breakfast?

If you answer no to any of these questions, you are cutting off one avenue of meaning, one avenue where significance may reside. You are cutting off connections; and meaning is the existence of significant connections.

It is inevitable that meaning and significance are lost. The passage of time compels it. Formerly, "esquire" designated a young, shield–bearing aspirant to knighthood; now, in the United States, it is a slightly silly honorific postfixed to the names of lawyers. Formerly, the Thanksgiving holiday was, literally, a giving of thanks to God for his bounty; today it is primarily a secular family holiday, an occasion for far–flung relatives to foregather and have a good meal. "To believe in the immortality of a poem," writes Octavio Paz, "would be to believe in the immortality of language. We must bow to the evidence: languages are born and die; any meaning will one day cease to have meaning."

Meanings may be lost irrevocably. We do not now know the meaning of the construction at Stonehenge. Cogent explanations have been proposed, but they are not accepted by all scholars. We do not know the manner in which the High Priest in Jerusalem pronounced the YHVH, nor the full import of its invocation.

The Extirpation of Meaning

When a person dies, so goes an old saying, a world of meanings is lost. Meaning is lost through the erosive effect of time, through inertia, through misadventure. It may also be lost through calculated violence. When books are burned and libraries destroyed, when censorship cuts and blots, meaning is lost. Such losses, brought about by deliberate and specific expungings, can be called the extirpation of meaning.

Iconoclasm—taken in the literal sense—is extirpation of meaning, and when the Early Church confronted the classic pagan world, much more was lost than a few bronze idols. When Henry VIII destroyed the monasteries of England, when Hitler instigated the Kristallnacht, meaning expired. The rewriting of history, wherever it occurs, the deliberate misrepresentations that result from cultural conflicts, as, e.g., between the contemporary Marxist and non–Marxist worlds, are attempts at wholesale extirpation of meaning.

Losses of meaning occur when intellectual ideals come into conflict. Such conflicts are most often centered around one question: which forms of knowledge are best or most important? In more virulent conflicts, the question becomes: which forms of knowledge are true and good, and which are false and evil? When one side is bludgeoned into submission, or, over a period of time, disappears by attrition, a massive loss of meaning may result. Such literature of the Graeco–Roman world as survives is principally that which was allowed to be read in schools. The letters of Symmachus (c. 400) contain a pathetic statement of how, in his day, the classics were no longer known, read, or copied.

There are many instances of ideals in conflict. One was the conflict in the 8th Century B.C. between rite and right (between Amos and Isaiah). At the time of Plato, the battle was between poetry and philosophy. In the 13th Century there was fought the "battle of the seven arts": of grammar, dialectic, rhetoric, geometry, arithmetic, astronomy, and music, which is the most fundamental? The battle between faith and reason has a long history, and, for the past several centuries, we have seen the battle between the scientific–technological and the humanistic cultures. In this battle, the former has, by and large, been the aggressor. Subjecting all intellectual material, including its own material, to a stringent methodology and value system, scientific culture has found the humanistic culture wanting. The view of Nobel Laureate Sir Francis Crick, quoted on page 252, displays the full fury of a crusading spirit. In Crick, materialism itself becomes the true religion. He is writing at the peak of public agreement that science has made this a better world and that the future lies with science.

Extirpation Through Intellectual Processes

The mind that thinks and creates can also be a mind that extirpates. St. Augustine was steeped in classic literature. After his conversion, he came to feel that it was dangerous for the public to read this literature. The holy deeds of the sainted Zacharias of the sixth century consisted simply in his going from village to village and burning books.

Let us see whether we can locate losses of meaning farther back in the intellectual process. Focusing on the mathematical mind, we have observed that it abstracts, generalizes, and dichotomizes. Consider abstraction. We observe that two fingers conjoined with two fingers make four fingers, and two cows conjoined with two cows make four cows.

When we proceed from these observations to the statement $2+2 = 4$, we have the simplest sort of mathematical abstraction. In the formulation of this arithmetical equation, we have, in the words of the dictionary, extracted out "the concentrated essence of a larger whole." Abstraction is the source of great benefit and also the source of possible damage. The damage derives from the self–deception that one has, indeed, discovered the essence of the larger whole. Abstraction is extraction, reduction, simplification, elimination. Such operations must entail some degree of falsification.

Now examine generalization. It works something like abstraction. Here is Jane and here is Dick. We may then discover that Dick and Jane are both instances of children, and that this notion of children contains some Dick–ness and some Jane–ness. If we then replace (the notion of) Dick and Jane by the notion of children and refocus our discussion accordingly, at this juncture a loss of specific meaning occurs.

I recall a taxi ride years ago from Gatlinburg, Tennessee, to Knoxville—a drive through the countryside of over an hour. The taxi driver was a garrulous fellow, and talked to me about local things; people, places, events. In my responses to him, I noticed that I often made generalizing conclusions or wrapups on the basis of what the driver had told me. The driver could not respond to these generalizations. He always went back to specific individuals. It struck me then with particular force that there was a profound gap between the professional way of thought of the so–called learned community, of which I was a member, and the rest of the world.[1]

Abstraction and generalization are two characteristic features of mathematical thinking, and mathematization is one of the crucial ways in which meaning is transformed and sometimes lost. If mathematics resided entirely in the mind and, although it derived from our interaction with the external world, if it played out all its themes and variations entirely in the mind, then it could do no great harm. But this is not the way it works. Mathematics comes from the interaction of the mind with the external world, and this interaction simultaneously creates mathematics and transforms our perceptions of the external world, and these then create new interactions. Unconscious mathematics resides equally in our monetary system and in the chip that controls a prosthetic device for the disabled.

When mathematics affects individual people, we must watch out and be careful, for the mathematics is often taken for granted; it is rarely questioned. Consider a specific individual named Jane Smith. This name tag carries, probabilistically speaking, a certain amount of information. Jane is female and is to be found in an English–speaking environment. In a mathematized version of Jane Smith, she might be listed as 072–33–2904, and this number tag would carry certain other information. This ID number might be an abbreviation for a whole dossier of information about Jane Smith. This information would have been obtained somehow and then digitalized. It might be true or might be false, in any case it can be only an incomplete representation of Jane who, in the last analysis, defies total verbalization and digitalization. The whole person is replaced by a part, her dossier.

The dossier 072–33–2904 might contain, for example, information on whether Jane Smith is a happy individual. Jane's supervisor at work might have answered an elaborate questionnaire designed by some well–trained and well–meaning graduate of a business school, well versed in psychology and in the statistical interpretation of questionnaires. This particular questionnaire might have contained the instruction: "Locate the subject on a 0 to 10 happiness scale." The potential benefits of this kind of thing are clear, as well as its concomitant evils.

When mathematics dichotomizes sharply, when it insists that the difference between a 7 and an 8 is absolute; when it carries this into society and says, for example, that those whose happiness number is below 8 are unqualified for a job (we have to draw the line somewhere, don't we?), then the wonderful Platonic precision of mathematical ideas can become a source of evil even as it confers benefits. *The very existence of the characteristic features of mathematics attracts the world and invites the world to apply them, come what may.*

While precision is a feature of mathematics, so is its opposite. Controlled blurring is a relative newcomer to the mathematical scene but is now commonplace. It goes by the name: statistics. Statistics (as opposed to mere data collection) begins when one agrees to form averages. Bill weighs 168 pounds, John weighs 190 pounds, and Bobby weighs 161 pounds. Their average weight is 173 pounds. This last statement is a compression of the first three. There is a loss of information and a loss of meaning in passing from the first three numbers to the fourth. There is, of course, a gain in the recognition of the empirical fact of the stability of averages. It may be that one of the reasons why probability and statistics did not take off until the 17th century was precisely the refusal of people to suffer the loss of the sense of the individual.

As a mathematical society, we are saturated with this kind of thing, and we have an expression that describes our malaise with it: we call it the dehumanization of the individual. The final intent of the application of mathematics to people is to be able to compare two individuals or groups of individuals, to be able to arrive at a precise and definitive opinion as to which is taller, smarter, richer, healthier, happier, more prolific, which is entitled to more goods and more prestige, and ultimately, when this weapon of thought is pushed to its logical limits and cruelly turned around, which is the most useless and hence the most disposable.

Whenever anyone writes down an equation that explicitly or implicitly alludes to an individual or a group of individuals, whether this be in economics, sociology, psychology, medicine, politics, demography, or military affairs, the possibility of dehumanization exits. Whenever we use computerization to proceed from formulas and algorithms to policy and to actions affecting humans, we stand open to good and to evil on a massive scale. *What is not often pointed out is that this dehumanization is intrinsic to the fundamental intellectual processes that are inherent in mathematics.*

Formalism

I once had dealings with a man who carried on his business from his apartment. He had a large clientele. He lived on the fourth floor of the apartment house, and access to his apartment was gained from a rear door on the ground floor adjacent to a parking lot. This door was normally locked.

Now, the manner of going up to his apartment was this. After calling him up just before your appointment, you parked your car in the lot and then honked your horn. The man would then come out onto his back porch and let down a key to the ground floor door on a long string. You would open the door with the key, find your way up to his apartment and give him back the key.

This procedure struck me as highly eccentric and mysterious. One day I asked him to explain what was going on.

"Why don't you simply have an electric lock installed so you could push a button in your apartment and it would open the bottom door?"

"There is an electric lock already on the door."

"Then for heaven's sake, why don't you use it?"

Construction in wood. Artist: Louise Nevelson. (*Collection of P.J.D.*)

"I'll tell you. About ten years ago, my wife and I were divorced. It was a nasty business, and for months and months after the divorce she would come around here causing me trouble. One day I simply decided that when someone pushed the button downstairs, I would ignore it, but of course I had to make provision for my clients. So there you are; it's not really mysterious at all."

"I take it she still comes around and tries to make trouble for you?"

"No. She died about five years ago!"

Formalism, in the sense in which I will use the term, is the condition wherein action has become separated from integrative meaning and takes place mindlessly along some preset direction. In the field of social psychology, some authors have called this "functional autonomy." In other areas the word formalism is used with slightly different but related meanings.

In mathematical philosophy, formalism is the position that regards mathematics as the study of formal deductive systems. Mathematical truth is simply provability in the system, and there is and can be no ultimate meaning to mathematics other than the operation of naked

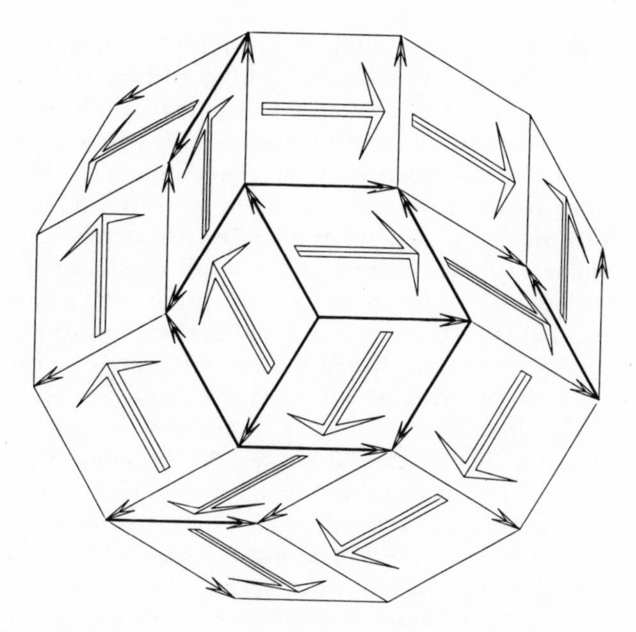

Orientations in six dimensions. (*Courtesy of Fred Bisshopp.*)

symbols according to fixed rules. This meaninglessness was stated in the already cited quip of Bertrand Russell, that "mathematics is the one subject in which we never know the meaning of what we say, nor whether what we say is true" (since the abstract mathematical form is only a carrier for all possible interpretations and meanings).

In literary criticism, formalism is the program which breaks down texts and creates a taxonomy of basic themata. In literary practice, formalism builds up new texts, largely on the basis of these themata.

Structuralism, as a school in literary theory, is allied to formalism. The following description is from M.H. Abrams' *A Glossary of Literary Terms.*

> "The following are among the salient features of much structuralist criticism.
>
> 1. In the structuralist view, a literary work is a mode of writing ('*écriture*'), constituted by a play of various elements according to purely literary conventions and codes; these factors within the literary institution generate literary 'effects,' without reference to a reality existing outside the system itself.

2. The individual author, or 'subject' is assigned no initiative, expressive intention, or design in producing a work of literature. Instead, the conscious 'self' is declared to be a construct that is itself the product of linguistic conventions about the use of the pronoun 'I;' the mind of an author is described as an imputed 'space' within which the impersonal system (or *langue*) of literary conventions, codes, and rules of combination gets precipitated into a particular written text.
3. In a similar fashion the reader, as a conscious, purposeful, and feeling individual, is dissolved into the impersonal activity of 'reading,' and what is read is not a text imbued with meanings, but ecriture, writing. The focus of structuralist criticism is on the activity of reading which, by bringing into play the requisite conventions and codes, makes literary sense—that is, endows with form and significance the sequence of words, phrases, and sentences which constitute a piece of literary writing. Structuralist critics earlier held that such reading, although it produces a plurality of meaningful effects rather than a single correct meaning, is nonetheless to some degree 'constrained' by the implicit codes of the inherited system. Recently, however, a group of more radical critics, with Jacques Derrida as a leading figure, have proposed that literary écriture provides only 'marks' which set off (or should set off) in the reader an unconstrained, or 'creative' play of unlimited significations."

In my mind's eye, I can trace the origins of this description through Russell and back to Euclid. Is it too naïve to hope that literary critics will ever again play Ann Landers to Anna Karenina? Or is Tolstoy's heroine merely an instantiation of certain formal kinship relations explicated after the manner of the abstract cultural anthropologist?

Formalism, in the sense in which I use it, has one of its earliest descriptions in the writings of Isaiah (c. 8th Century B.C.). Isaiah objects that the act of fasting has become separated from the ethical meaning of the act:

"Is it a fast like this that I require,
a day of mortification such as this,
that a man should bow his head
like a bulrush
and make his bed on sackcloth and ashes?
Is this what you call a fast,
a day acceptable to the Lord?
Is not this what I require of you as a fast:
to loose the fetters of injustice
to untie the knots of the yoke,

to snap every yoke
and set free those who have been crushed?"[2]

Formalism is the traffic light in a residential area that goes through its color changes long after the populace is asleep.

Formalism is when your pet dog turns around so as to ward off enemies before it curls up on the living room rug.

Formalism is when seven year olds compete in a spelling bee, having boned up on hard words whose meanings they do not know.

Formalism is when nine year olds go through the arithmetic steps in 951×202 or $951 - 202$, in the absence of any quantitative feeling for the numbers involved.

At a higher level, my wife took a course in college given by a world renowned logician. She got an A in the course. Yet, years later, she confessed to me that she hardly knew what had been going on. She had simply learned how to reproduce formal manipulations well enough to pull down an A.

Stella Baruk, a seasoned mathematics teacher in the French school system, writes: "From Pythagoras in antiquity to Bourbaki in our own day, there has been maintained a tradition of instruction—religion which sacrifices free understanding to the recitation of formal and ritual catechisms, which creates docility and which only simulates sense. All this has gone on while the High Priests of the subject stand in their corners and laugh."

But some of the High Priests of mathematics must weep, because it is in the very nature of mathematics that its abstract symbols suffer a loss of meaning even as they gain in generality. Over the ages, mathematicians have struggled to restore thought and meaning to mathematics instruction, to provide alternatives to the formal and ritualistic mode of learning in most mathematics classrooms, but in spite of new theories, new applications, new courses, new instruments, the battle is never won. The fight against formalized, unthinking action is perpetual.

Formalism is a medical researcher who uses a computer to calculate standard deviations in an experiment that was ill conceived and ill executed. The statistical information is put into the paper simply because in today's research world it is expected. It is part of the credentiation process.

Formalism is the piling up of money beyond what an individual, his family, or heirs can consume. It is the reduction of the purpose of a company to one motive: to show a profit.[3]

A rich man I know once told me that he kept on manipulating in business (and piling up money) because it was the only way he knew of "escaping from the real world." Where is the real world located?

"Textron finds itself a target for buyout by Chicago Pacific." So read a headline in the Providence Journal of October 25, 1984. Now, Textron itself was one of the first conglomerates. It has bought up other companies for over fifty years. So here is the modern business mind to be seen in all its glory, dedicated to creating and manipulating abstractions and abstractions of abstractions. As concepts are converted to actions, and as people are hired, fired and shuffled, the material products at the bottom hardly improve. While one can see why the current structures of business and of law lead to such amalgamations, it would be hard, indeed, to demonstrate a net benefit to our civilization. Formalism is running wild.[4]

Formalism is a disproportion between form and content: a euphuistic style, overblown baroque decoration, a bit of junk food frosted and adorned but with no nutritive value.

"If form predominates," writes Sir Kenneth Clark, "there is a loss of vitality and of that humanity which should underlie even the most idealized construction. If subject predominates, the mind releases its hold. In both cases, the chance of a masterpiece is diminished." (*Idea of a Masterpiece*, p.29).

I have a short wave radio receiver and occasionally tune in to the amateur band and listen to the hams. What do hams talk about to one another? That they drove to Schenectady to visit Uncle Harry? That they are angry because the shenanigans of the mayor of their city are beyond belief? Or are they transmitting a message to the Coast Guard to pick up a disabled sailboat? No. None of the above. Ninety–nine percent of the conversations of hams relate to the technical problems of sending and receiving messages and of the comparative merits of equipment.

In the same way, it can happen that the issues of war relate only to the way the war is fought:

> "The British historian E.P. Thompson has pointed out, in his book *Beyond the Cold War*, by conditioning military and political elites, on both sides, to act in accord with the first premise of adversary posture—to seek ceaselessly for advantage and to expect annihilating attack upon the first sign of weakness—strategic doctrine could tempt one side (if a manifest advantage should arise) to behave as theory prescribes and to seize the opportunity for a preemptive strike. And what would the

war, then, have been *about*? It would have been about fulfilling a theorem in deterrence theory. It's a striking historical fact, one that should make us reflect, that the severest crisis of the nuclear age, the Cuban missile crisis, was *about* the weapons themselves."—J. Schell, The New Yorker, January 4, 1984.

Now, I am aware that there are no pure tendencies, there can never be a complete separation of meaning and action. We all understand the reason why the traffic light is blinking when there are no cars. We know that it can be said that an act has meaning precisely because it has no meaning; or that anarchists have sometime carried out "meaningless" actions precisely to shock an indifferent populace into creating meaning.

In using the word "formalism" and in pointing to instances of it, I do not mean to suggest that formalism is a material substance that resides in the atmosphere and can infect us; nor do I suppose that I accomplish a great deal, for example, by identifying as an instance of formalism what is usually called greed. I use the term formalism only in order to call attention to a natural tendency of form and function to get out of balance or to part ways. We cannot prevent their separation and probably should not try to do so, but we should be aware of the process, so that we may institute countermeasures when it gets out of hand.

Disaster Through Intellectual Processes?

Even more influential than soldiers in shaping America's weapons policy has been an elite group of economists, mathematicians, and political scientists who, beginning in the 1950s, preempted the bomb as their special intellectual property, established themselves as a proprietary priesthood, and sought to impose logic on inherently irrational nuclear conflict.

The bomb offered them a unique chance for theorizing free from empirical challenge, for, although the practices and doctrines of conventional warfare have been tested in a thousand battles, the conduct and consequences of a nuclear exchange are, and must remain, pure speculation—unless and until such an exchange blows up large parts of the world including the intellectual speculators. Thus the theoreticians did only what came naturally to them; they saw an unprecedented opportunity to deal in the realm of pure abstraction unmenaced by experience. Because no one has seen either an angel or a nuclear exchange, the hackneyed analogy to medieval scholastics is by no means farfetched.—George W. Ball[5]

Human suffering must not be abstracted. This should be the first law of ethics, the Golden Rule.

An absurd commandment, when you think about it. If it could conceivably be carried out, it would destroy, in one blow, all applications of mathematics to the human sphere. Gone would be money, economics, laws of damages, insurance, operations research, statistics, medicine, social planning, military technology, and strategy. Our lives would be primitive and naïve, unrecognizable.

In my mind it is no accident that the great evils of the period 1933–1945 were perpetrated in a country that was the world leader in theoretical science and mathematics. It was not necessary for the policy makers to have understood mathematics; it sufficed that a certain spirit—part of which was mathematical—was in the air. Numbers, tattooed on the arms of the victims, reduced them to the level of branded cattle.

There is an image, part mythical, part parodic, of the mathematician as the man who is all brains and no heart, all precision, all program. Isolated from society and driven by infantile fantasies, he is the mad professor, if you will. This image indeed expresses a truth about the relation between society and mathematical methodology, though not, of course, about specific mathematicians. Do not get me wrong. I am not suggesting that a high degree of mathematization inevitably leads to a holocaust; only that the spirit of abstraction and the spirit of compassion are often antithetical.

Is there such a thing as historical explanation, or is there only exposition and an invitation for the reader to accept the exposition as explanation?[6] "Sequential history" moves forward in time, through the narration of event.[7] "Explanatory history" moves backward in time. "Interpretive history" explains the relationship between individuals and groups of individuals, their institutions, their frames of mind, the role of the culture. History as pure narrative makes no sense. You may learn about all the events that led to the destruction of Rome, and still you will be unsatisfied. What was behind it all? An interpretation is required. And if you are told, (as in Rostovtzeff, *The Social and Economic History of the Roman Empire*, p. 486) that men's minds turned gradually from the things of this world to thoughts of heavenly salvation, then the relationship between this explanation and actual events must be made palpable.

If the major unsolved problem of the history of Western civilization is to account for the collapse of the Roman Empire, then surely the major problem of contemporary history is to account for the Holocaust. The narrative aspects of the events in Germany during 1933–1945 are still being assembled. Alongside the narratives have been many attempts at interpretive histories. Such interpretations have been organized along a few dominant themes. Jung suggested the resurgence of the "Wotan Archetype." Wilhelm Reich suggested the suppression of genital sexuality. Jean Paul Sartre suggested intellectual jealousy. Erich Fromm pointed to the desire to control, and to the necrophilism of Hitler. Elias Canetti suggested that the German inflation of the early 1920s, which introduced huge, unreal numbers, disturbed the relationship between the abstract and the concrete. George Steiner suggested that Jewish monotheism, Christian piety, and Marxist messianism set perfectionist goals which mankind found impossible to achieve, and that the Holocaust was a violent reaction against these ideals.

I will add one more vision of perfection to Steiner's list: the Greek idea of a perfect truth attainable through mathematical abstraction. I should like to suggest that advanced mathematization, through abstraction and subsequent loss of meaning, played a role. It is a possibility that merits the collection of evidence, merits speculation and argumentation; for, of course, the full story does not involve Germany alone, nor does it stop with the events of 1933–1945.

It cannot be denied that World War II took place in an era of high mathematization. Mathematicians as *such* were employed in the development of codes, in operations research, in ballistic control, in radar, in atomic energy. Undergraduates now learn that this war was the genesis of many significant theories in applied mathematics. Military technology and strategy today have so much mathematical basis that we see the grim truth of a saying that was common in the late 1940s: whereas World War I was a chemists' war, and World War II was a physicists' war, World War III will be a mathematicians' war. (Let us hope that the whole thing is fought abstractly, inside the memory of a supercomputer!) In this connection, consider Edmund Wilson's diary entry in 1944:

> It may be that one thing which is responsible for the war is simply the desire to use aviation destructively. It must be a temptation to humanity to blow up whole cities from the air without getting hit or burnt oneself, and while soaring serenely above them. Many must feel vicariously as I do the thrill of doing this—I felt it when the Germans were bombing London before we had begun bombing them. It is the thrill of the liberation of some impulse to wreck and to kill on a gigantic scale without

> caring and while remaining invulnerable oneself. Boy with a slingshot shooting birds—can't help trying it out. This is true of mechanical warfare in general: the guns batter down at long distance, the tanks flatten out without feeling, the planes wipe out whole cities without one's having to picture what has happened. It is the gratification of the destructive spirit, as it were in a pure *abstract* form: if the enemy does not get you, you are free as a bird of the consequences. You are further removed from your object by the intervention of the mechanical bombsight: your end of the process is *mathematical,* the machinery sends the bomb to its target. Constraints of the conventions and codes that we live under, and that we are glad to see smashed.[8] (Italics mine)

A few months separated the Holocaust in Germany from the Holocausts in Hiroshima and Nagasaki. The atom bomb was made possible by the same talented, theoretically oriented individuals who had been driven out of Europe. A computation was made: drop the bombs and lives will be saved thereby. What other criterion can there be in times of national peril and international madness? The subsequent bewilderment of exiled and native intellectuals who had believed that the cultivation of theoretical studies was the hallmark of a society that was pure and good; the sense of sin experienced by J. Robert Oppenheimer when, bound to the rocks like a new Prometheus, the eagles of guilt and obloquy plucked at his liver; all these led slowly to an altered opinion that if, indeed, science can remake the world, it must not be given free rein when it asks to do so.

The physical world has not yet been revealed to us in all its features. If it had been, there would be no need to do any more experiments. Nor do we as yet fully know the world of the mind and its relationship to the physical world. It might be thought that to assert that the ills of the age derive from something as basic as abstraction is to assert nothing, really. After all, abstraction can work in many ways; it is everywhere. It would be like blaming the human condition on gravity, on the fact that things fall when you let go of them. This much can be said, however. Early on, mankind learned to avoid the edge of cliffs. In the automobile age, it has learned not to drive at high speed over potholes. *We must learn as well what to avoid in the landscape of the mind.*

Loss of Meaning Through Computers

Not long ago, I was invited to participate in a debate on the use of computers in mathematical education. Are they a miracle or are they a menace? The organizers of the debate asked me to present the "menace" position. I agreed, although I warned them in advance that actually I thought that they were both.

At the time the invitation came, I had just finished reading an essay by Clifford Truesdell entitled "The Computer: Ruin of Science and Threat to Mankind." I was then giving a course on the nature of mathematics, designed for undergraduates who intended to go into the teaching of secondary mathematics. Now, Truesdell is a world–renowned applied mathematician and historian of science. I concluded that by combining the thoughts of Truesdell, who looks at the matter from the point of view of the evolution of scientific ideas, with those of my class who look at the matter from the point of view of pupil–teacher interaction, I should find plenty of arguments, and would hardly have to write my own debate.

The most trenchant of Truesdell's arguments is that the computer, as a research tool, fosters "floating" mathematical models:

> The models rational mechanics provides are strictly logical; they are *deductive models*, articulations of a particular theory. Classic science embraces also *inductive models*, summarizing an organized body of experimental data. Models of both these kinds are systematic. They teach us to find structure in experience, not merely to imitate one or another detail.
>
> Recent research resorts more and more to *floating models*, which treat phenomena severally, with no subsumption under general theory or organized knowledge gained from experiment.[9]

As recent examples of floating models, Truesdell cites applied catastrophe theory and soliton theory, directing his harshest words at the former. The next decade will decide whether or not these are good examples of the phenomenon he has isolated. There is no doubt, however, that the computer does bring a new spirit into science. By turning attention away from underlying physical mechanisms and towards the possibility of once–for–all algorithmization, it encourages the feeling that the purpose of computation is to spare mankind of the necessity of thinking deeply. By calling in standardized paradigms, it is imagined, the computer will be able to formulate its own automatized theories of what is happening.

Hannah Arendt describes the degradation this way:

> Most commentators see technology, science, as an instrument for the self–assertion of reason, which, itself needs no additional justification.

> Modern times, dominated by technology, are characterized precisely by the fact that reason, in the sense of an originally given self–revealing contemplative understanding, is *lost*, and is replaced by a detached (technological), actively preoccupied with abstract mathematical theory and physical replication.[10]

When confronted with the prospects of widespread use of the computer in secondary education, what reservations did my aspiring teachers express? The very first reservation derived from their knowledge that students—from those in elementary mathematics to those in calculus—resort more and more to their hand–held computers to do such things as 8 × 6. (I do not joke: this was confirmed by the experience of Professor John D. Neff of Georgia Institute of Technology.) We are confronted here with the loss of a fundamental base of memorized or internalized knowledge. Also of concern to them was the decreased social interaction in both the scientific and humanistic areas of education, and in particular, the loss of human role models in the mathematical process.

There is a myth of the isolated scientist, of an Einstein who is holed

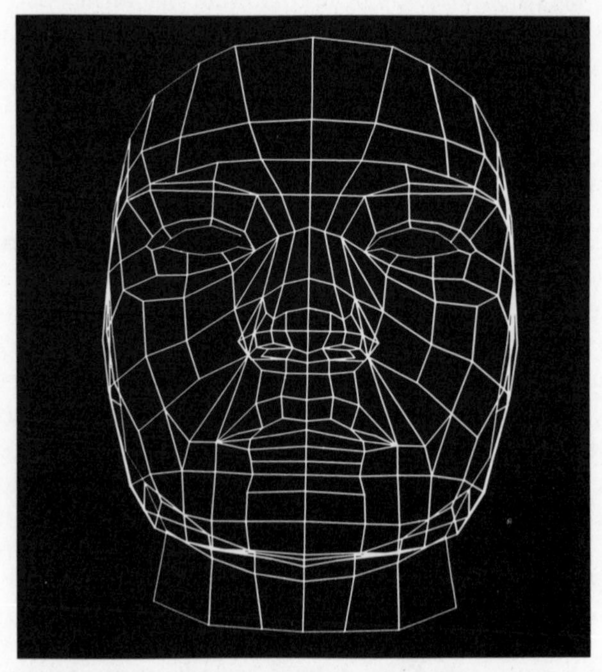

Human head geometrized. (*Courtesy of the Lownes Collection, John Hay Library, Brown University.*)

up in a lighthouse with his books and stars, who there, fed by the seagulls, produces the general theory of relativity. This picture is false. Scientific meaning derives from both social interaction and interaction with the physical world through experiment. The strictures both of Truesdell and of my class of undergraduates can be summed up by saying that the computer will lead to an excess of formalism, wherein symbols are separated from any underlying meaning and become the sole arbiters of their own meaning.

Mathematics and its child, computer science, look for *total solutions.* In mathematics, a total solution is one in which a problem has been so thoroughly analyzed that all instances of it may henceforward be solved by rote, by an algorithm. In computer science, a total solution is a system that enmeshes our entire life in an automatized network, in which the human role has been completely usurped by the formal transformation of symbols.

The Dream of Descartes began with the possibility of automating geometry. It was inevitable that one would think next of the possibility of automating all thought and judgement. One then advanced to the proposition that natural language is computation, that vision and the other senses are computation, and that emotion is computation. *Computo, ergo sum* seems to be the ultimate expression of the Cartesian insight. It abandons humanity, replacing it by an abstract surrogate.

The Rebirth of Meaning

The main philosophical problem of our generation is that of reestablishing meaning.

The experience I am about to describe happened to me several times at the beginning of my career. Most people, I am sure, have gone through something similar. As a professor of mathematics, most of my teaching takes place in lectures. Very often, I work at the blackboard with my back to the class. As I am working along, talking, explaining, writing strings of symbols, quite suddenly the symbols give way and become vague in my mind. They lose coherence. They lose their relation to each other and to what I have been saying. Their meaning has drained away and they stand on the blackboard in front of me as just so many strange and naked geometrical shapes.

At this point in my lecture I am stuck. After a moment of floundering around, I come to a complete halt. I am embarrassed, and my class is embarrassed by my silence. I break out in a sweat. I try to reestablish the meaning of what I was doing, but it will not emerge. I back up and review in my mind what I had been saying up to that moment of crisis, but, although I can replay the words, the words seem empty. I go on to another subject. I say to my class, feeling guilty as I do, that I will really clarify matters the next time we meet.

What is going on here? A failure of memory? Stage fright? One reads that even the most seasoned performers can be victims of stage fright. Does the crisis have a physiological basis? Has the ventilation in the classroom been poor? Possibly. In any event, I relate these episodes, episodes that have been repeated in dreams of nightmarish intensity, for use as a guide to larger, much more significant crises of meaning.

Return with me to the classroom. Of course, I lecture from prepared notes—occasionally from a textbook. (Some teachers do not, thinking that it interferes with the intensity or the authenticity of the performance.) When I got stuck, I could, if I had wanted to, simply have recited what was written in my notes. I could have droned away, faking it, and hoping that my students would not catch on to the fact that my performance was devoid of those qualities that form the basis of teaching; the understanding, the personal engagement and experience, the insights, which, taken all together, impart meaning.

Early in the game, I found that I had to prepare notes on two levels. The first level, the formal level, contains the material that is part of the public record. The second level contains a thin sprinkling of private notes to myself, interpretable only by myself, which helps me establish the meaning of what is written on the first level. Armed with such a set of notes, I find I can, whenever the meaning or whenever the formal aspects weaken as I lecture, reestablish either one with reasonable ease. Forms collapse, but they can be rebuilt. Meaning erodes, but it can be reestablished.

□□
□□

I am in the middle of doing something. Perhaps it is driving to Boston. Perhaps it is mowing the lawn or listening to a Schubert symphony. Or the telephone rings and I am about to pick up the receiver. Suddenly, in my mind, the act becomes weakened and gives way. I have driven to Boston many times. Why should I be doing it now? What conceivable difference would it make to me or to anyone if I did or did not keep my engagement? I have heard Schubert's symphony many, many times. Yet another time? The purpose, the importance, the meaning of my actions drain away and leave me with naked, formal, unexplainable motions which, once initiated, I carry out because I cannot find sufficient reason to terminate them. In such a crisis, I am at the very least, like Buridan's ass which, when it came to a fork in the road, couldn't find sufficient reason to go either right or left and so could do neither; or, at the worst, like the Sisyphus of Albert Camus who does not terminate by suicide his cyclically endless labor and suffering simply because of his sense of the absurd.

A true story. My mother, dying, washed out a pair of diapers for her infant grandchild. Think about that when next you tell the Sisyphus story.

Meaning reestablishes itself. I know it does because I have experienced it happening over and over again. A more complete description would be to say that *meaning continuously drains away and continuously reestablishes itself.*

I would not be telling the truth if I said that I have a formula for composing a two level set of notes, one level for action and one level for meaning, which is guaranteed to get me rapidly out of an existential jam. I do not. I am not even clear in my own mind whether the division of the world into action and meaning, like other famous philosophical splits (mind and matter, form and substance, continuity and discontinuity), is part of the objective world or is only an aspect of how our western brains create and use language.

I do not assert there cannot be such a formula; only that I have none.

I am aware of the ingredients out of which meaning is created. There are many such: love and language, myth, rational thought and irrational impulse, human institutions, law, history, duty, ritual, religious faith, the mystic, the transcendental, the allegorical, the aesthetic sense, play, the world as a puzzle, the world as a stage, the contemplation of life and death, the necessities imposed by physics and biology; all of these and hundreds more are avenues to meaning. *Nor should we be too hasty in closing off any such avenues, for they are precious.*

I know, for example, that among scientists of the past four or five generations there has been an arrogation and an abrogation of meaning.

A true story. I know a famous astronomer. Call him Ichabod. He is never so happy as when he is contemplating and explaining the movements of the stars. He would certainly assert the first part of the verse

"The Heavens declare the glory . . . ."

But he couldn't (or wouldn't) finish the verse. "Do not fill your mind with that kind of trash. It is nonsense. Meaningless concepts. Counterproductive." Now, a strange thing happened. You might say it was inevitable that it happened. As time went on, more and more astronomy was discovered. Strange particles from high energy physics. Strange phenomena from radio astronomy. The cosmos grew complex and the simple conceptualizations and unifications of previous decades lay shattered about him. He was, of course, part of the process. Gradually, in his mind, the meaning of the enterprise drained away and it all stood before him as mere, naked, existence.

When last I spoke to Ichabod, the whole of the astronomical cosmos was unending and meaningless, ununderstandable, and purposeless; not even a joke. His own role in it was devalued. The Glory had departed.

"Tell me, O moon, what worth
Be this life to the shepherd, or yours to you?
Where does this, my brief roaming, tend? And where
Your everlasting course."[11]

The famous philosopher Sir Karl Popper has stated baldly: history has no meaning. But he doesn't leave it there. He goes on to assert that *we are able to give it meaning.* That's the thing: meaning can be established.

□□
□□

I am lying in bed. The clock says it is time for me to get up. My affairs are a mess. The world is a terrible place and is poised at the brink. Why should I get up? Simply because this world is the only world we have? I move the first two joints of my index finger. Gee, they work! Maybe all my fingers will flex. They do. Maybe I can move my arm. I can. It even feels good to do so. And so, after a minute, I am up and dressing. It is as though the potential for existence has elicited, of its own, the meaning for such existence. To assert this and nothing more is to advocate a philosophy of formalism, where the sole meaning of an act lies in the act itself. I know that, in my case, I need all the traditional avenues to meaning to get me through the day. They assemble themselves even though I have no formula or strategy for their combination.

Given a chance, meaning will establish itself. It must, else there is no possibility for sanity and coherence. Simultaneously, meaning must decay, else there is no possibility of growth. The meaning of a friendship at the age of thirty must be different than what it is at the age of sixty. What is crucial is the balance. I am constantly fighting to establish a balance. What is more, I am aware of the struggle.

Notes

1. Cf. a description of the way Franklin Delano Roosevelt's mind worked:
   Possessed of an intellect that was broad but shallow, he collected facts and ideas as he did stamps and naval prints, letting them lie flat, distinct, separate in his mind, never attempting to combine them into any holistic truth. Indeed, he shied away from generalized thinking and abstract ideas. If never openly contemptuous of pure thought (certainly he was never assertively so), he had nothing to do with it personally, feeling it to be not merely irrelevant to his vital concerns but even hazardous to them insofar as it might distract his attention from small but important signs or cues presented him by and through his immediate environmental situation.—K.S. Davis: FDR as a Biographer's Problem. The Key Reporter, Autumn 1984.
2. Isaiah 58.5. Translation: *New English Bible*, Oxford University Press, 1971.
3. cf. Mark Schorer's assessment of Moll Flanders:
   . . . she has no moral being, nor has the book any moral life. Everything is external. Everything can be weighed, measured, handled, paid for in gold, or expiated by a prison term. To this, the whole texture of the novel testifies: the bolts of goods, the inventories, the itemized accounts, the landlady's bills, the lists, the ledgers: all this, which taken together comprises what we call Defoe's method of circumstantial realism.
   He did not come upon that method by any deliberation: it represents precisely his own world of value, the importance of external circumstance to Defoe.—Technique as Discovery, Hudson Review, Vol.1, No. 1, 1948.
4. In a widely quoted column, Russell Baker wrote in 1969:
   "It is not surprising that modern children tend to look blank and dispirited when informed that they will someday have to 'go to work and make a living.' The problem of course, is that they cannot visualize what work is in corporate America. . . . No so long ago . . . when a child asked, 'What kind of work do you do, Daddy?' his father could answer in terms that a child could come to grips with. 'I fix steam engines.' 'I make horse collars.' Well, a few fathers still fix steam engines and build tables, but most do not. Nowadays, most fathers sit in glass buildings doing things that are absolutely incomprehensible to children. The answers they give when asked, 'What kind of work do you do, Daddy?' are likely to be utterly mystifying for a child. 'I sell space.' 'I do market research.' 'I am a data processor.' 'I am in public relations.' 'I am a systems analyst.' Such explanations must seem nonsense to a child. How can he possibly envision anyone analyzing a system or processing a datum? Even grown men who do market research have trouble visualizing what a public relations man does with his day, and it is a safe bet that the average systems analyst is as baffled about what a

space salesman does at the shop as the average space salesman is about the tools needed to analyze a system.—"The New York Times. 16 October 1969".

5. Review of "Deadly Gambits": by Strobe Talbot in New York Review of Books, November 8, 1984.
6. cf. the classic line of Ring Lardner: " 'Shut up', he explained."
7. These terms are taken from Maurice Mandelbaum: *The Anatomy of Historical Knowledge*, Johns Hopkins University Press, Baltimore, 1977.
8. Edmund Wilson, *The Forties*, Farrar, Strauss and Giroux, New York, p.45.
9. C. Truesdell: Springer-Verlag, 1984.
10. The Human Condition, V. pp. 248–304.
11. Giacomo Leopardi, Transl. O.M. Casale.

Further Readings. See Bibliography

H. Arendt; R. Avenhaus and R. Huber; S. Baruk; G. Boynton; S. Bremer; K. Clark (1979); F. Crick; J. Ellul; R. Huber; R. Iano; C. Truesdell

VII
ENVOI

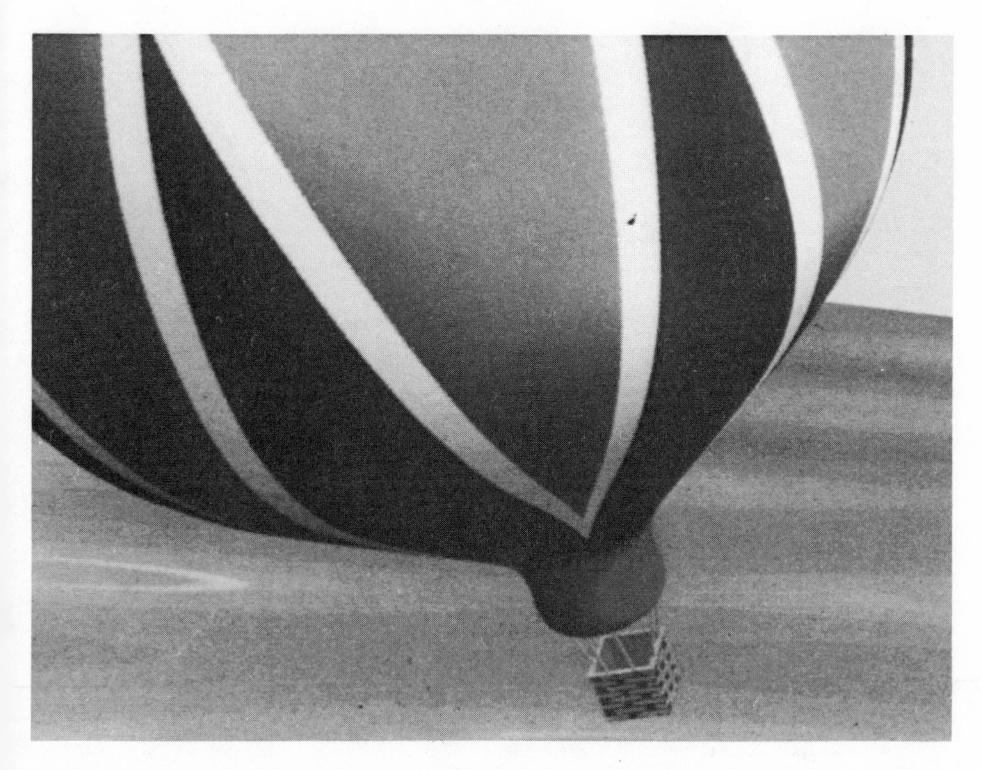

Balloon. Artist: Kelan Putbrese. (*Courtesy of Brown University Computer Science Laboratories.*)

Quod vitae sectabimur iter.
What path shall we take in life?

This book began with two quotations. The first is from Descartes and says that all intellectual matters can and should be unified by mathematization. The second is from Vico and says that mathematics is alienated from the human spirit. The human spirit resides in human institutions, and the human institutions, according to Vico, are the family, law, government, myth, religion, language, art, poetry, song.

There is a profound gap between these statements, and there is a profound gap between the types of world that each statement implies. The Spirit of the Age has been that of Descartes. Mathematization has been increasing and is still doing so. Its peak is not in sight. The human institutions in the sense of Vico are changing, eroding.

As the decades accumulate to centuries, will the Cartesian spirit also change? What will replace it? What will be the New Tendency? Who knows? But since the Vicovian institutions will be around for a long time, and since science and mathematics will also be around, one would hope that Cartesianism will be replaced by a fusion of the outlooks of Descartes and of Vico.

The computerization of the world represents an advanced stage of Cartesianism. Within that stage, programs become autonomous. We have even been given intimations of automated concept formulation and of action instigated as a consequence of such automation. If this indeed becomes the case, then *mathematics–computers really ceases to be a human institution.*

If the computer succeeds in splitting mathematics and science into two camps, the traditional classical camp and the camp of computers, and I think it much too early to tell about this, then the former will pass into the desiccated irrelevance of a classic, isolated from all its own self–generated concerns and clinging to a philosophy that pronounces its material to be apart from human origin and experience. The latter will pass into a state of "mechanized petrifaction embellished with convulsive self–importance" (these are the words of Max Weber, describing not computing, naturally, but advanced capitalism). In neither camp do we sense that mathematics is a human institution.

Mankind must achieve a synthesis. That such things have occurred in history is clear. One can point to the synthesis of the Christian World and the Classical World that occurred in the 12th–13th Centuries of which St. Thomas Aquinas was the outstanding spokesman. Within the current mathematical scene, although most signs point in the opposite direction, there are a few signs that are hopeful.

There is a reformulation of the major concerns of the philosophy of mathematics away from the logical and towards the phenomenological–historical–experiential. This takes mathematics "down from the sky," so to speak, and says, in effect, *we* are thinking these thoughts, *we* are writing these symbols, *we* are doing these mathematical things, and as a result of *our* activity, the consequences to *us* are such and such. Part of mathematics is beyond our control, just as part of the universe is. We may not be able entirely to separate the parts. Nonetheless a heightened sense of the interaction between mathematical thoughts and arguments and human life is emerging. Whether or not one has more or less mathematization may not really be the issue. What may be crucial is whether society develops a self–awareness that in its ordinary mathematical usages it is arranging itself in certain ways and hence is doing something to itself. In this way, mathematics becomes a human institution.

There is emerging a realization that some of the intellectual concomitants of mathematics are not really that far removed from similar components of the Vicovian institutions. Metaphor and analogy exist in mathematics and physics as well as in poetry and in religion. Rhetoric exists in mathematics (despite claims to the contrary) as it exists in politics. Aesthetic judgment exists in mathematics as it exists in the graphical or performing arts.

In its origin, in its development, mathematics requires full association with all types of human activity, mental and physical.

Though scientific histories of the past two hundred years have ignored or have been reluctant to admit the association, mathematics has drawn inspiration and nourishment from business, from religion, from law, from war, from politics, from ethics, from gambling, from metaphysics, from mysticism, from ritual, from play (look what a mathematical thing the children's game of hopscotch is), and not just from a "sanitized" physical science approved by positivism.

Mathematics has taken from these human institutions and has returned more than a tithe to them.

Man–machine symbiosis is clearly achievable — whether the machine is of the traditional mechanical variety or of computer type — and can be considered a fact of the evolutionary–adaptation process. The primitive bush is itself a vast and complex milieu, and when man came out of the bush he simultaneously created and adapted to new arrangements, some physical, some conceptual, which he perceived to be of his own making or which he asserted to have supernatural origins. Out of this adaptation came human institutions. The man–computer combination is the most recent phase of this process, and out of it can come human institutions of a new sort, institutions which, in every way, can be considered as natural as the traditional ones, and which can be judged by all the criteria used in judging the traditional ones.

All of these I believe are hopeful signs. One might imagine that in the year 2500, which should be well into the Age of the New Tendency, the shade of Vico will look down from Elysium with some surprise and considerable pleasure and say, "Mathematics is also among the human institutions."

If a synthesis cannot be achieved, if it comes to a showdown between man and mathematical science, then man would be best advised to stop the process. Let it fall into decay, just as a vast army of past activities have fallen into decay. The beautiful structures of thought will have proven, in the long run, to be pernicious; a very bad path for Descartes or humanity to have pursued with revelatory intensity.

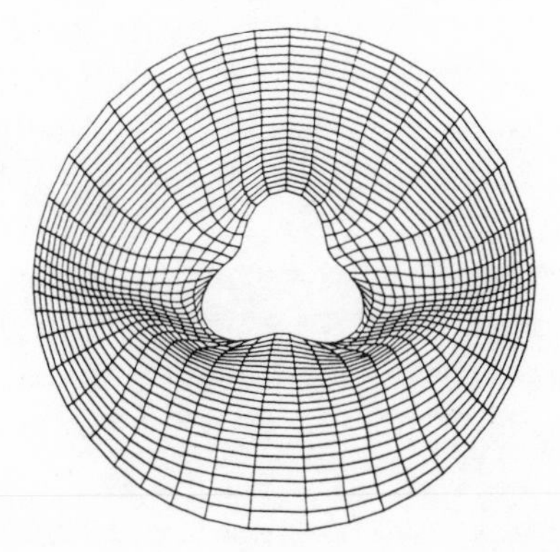

Adaptive grid for the solution of reaction diffusion equation. (*Courtesy of Jeremiah U. Brackbill.*)

Mathematics is changing and its role is changing. Those who entered the profession before the onslaught of digital computers are old enough to feel this change most profoundly. The mission of this book is to point to the necessity, as that change occurs, of developing a heightened awareness of the relationship between humans and the mathematics they have created. This awareness is necessary to shield us from the effects of the revolutionary waves of symbols that are about to wash over us.

Bibliography

AABOE, ASGER: Episodes from the Early History of Mathematics. New York: Random House, 1964

ADCOCK, CRAIG: "Conventionalism in Henri Poincaré and Marcel Duchamp." *Art Journal*, College Art Association in America, Fall 1984, 249-258

ALBERS, D.A. and ALEXANDERSON, G.L., eds.: *Mathematical People. Profiles and Interviews.* Cambridge: Birkhäuser, 1985

ALLEN, ALLEN: Some Necessary Conditions for Consciousness, Memory as a Sixth Sense, and Related matters. *Correspondence, Speculations in Science and Technology*, 4:441-445, 1981

ANDRESKI, S.: *Social Sciences as Sorcery.* New York: St. Martin's Press, 1972

ARBIB, MICHAEL A.: *The Metaphorical Brain.* New York: Wiley-Interscience 1972

ARBIB, MICHAEL A. and CAPLAN, DAVID: Neurolinguistics must be Computational. *The Behavioral and Brain Sciences*, 2:449-483, 1979

ARDEN, BRUCE W., ed.: *What Can Be Automated? The Computer Science and Engineering Research Study (COSERS).* Cambridge: MIT Press, 1980

ARENDT, HANNAH: *The Human Condition.* Chicago: University of Chicago Press, 1958. Particularly Chapter VI

ARIS, RUTHERFORD, and PENN, MISCHA: The Mere Notion of a Model. *Mathematical Modeling*, 1:1-12, 1980

(PSEUDO:)ARISTOTLE: *Traite du Ciel.* Traduction et Notes par J. Tricot. Paris: Libraire Philosophique J. Vrin, 1949

ARNOLD, V.I.: *Ordinary Differential Equations.* Cambridge: MIT Press, 1973

AVENHAUS, R., and HUBER, R.K. (eds.): *Quantitative Assessment in Arms Control.* New York: Plenum, 1984

AYER, A.J.: *Language, Truth and Logic.* Revised edition. London: Gollancz, 1964

BARBEAU, E.J.: Euler Subdues a Very Obstreperous Series. *American Math. Monthly*, 86:356-371, 1979

BARUK, S.: *L'age du capitaine. De l'erreur en mathematiques.* Paris: Editions du Seuil, 1985

BARZUN, J.: *A Stroll with William James.* New York: Harper and Row, 1983

BECKER, GARY S.: *A Treatise on the Family.* Cambridge: Harvard Univ. Press, 1981

BELL, ERIC TEMPLE: *The Search for Truth.* New York: Reynal and Hitchcock, 1934

BENNETT, CHARLES H.: On Random and Hard-to-Describe Numbers. *IBM Thomas J. Watson Research Center Research Report RC 7483*, 1979

BERGGREN, DOUGLAS: The Use and Abuse of Metaphor. *The Review of Metaphysics*, XVI:237-258, 450-472, 1962

BERLIN, ISAIAH: *Vico and Herder: Two Studies in the History of Ideas.* New York: Viking Press, 1976

BERLIN, ISAIAH: *Against the Current, Essays on the History of Ideas.* New York: Viking Press, 1980

BERLINSKI, DAVID: *On Systems Analysis.* Cambridge: MIT Press, 1976

BERLINSKI, DAVID: *Black Mischief.* New York: William Morrow, 1986

BLEDSOE, W.W. and LOVELAND, D.W., eds.: *Automated Theorem Proving: After 25 Years.* Providence: American Mathematical Society, 1984

BOOLOS, GEORGE and JEFFREY, RICHARD: *Computability and Logic.* Cambridge: Cambridge Univ. Press, 1974

BORGES, J. L.: Tlon, Uqbar, Orbis Tertius. *Labyrinths: Selected Stories.* New York: New Directions, 1962

BORST, C.V., ed.: *The Mind/Brain Identity Theory.* New York: Macmillan, 1970

BOYER, CARL B.: *A History of Mathematics.* New York: John Wiley & Sons, 1968

BOYNTON, G.R.: *Mathematical Thinking about Politics.* New York: Longman, 1979

BRAMS, STEVEN J.: *Superior Beings.* New York: Springer-Verlag, 1983

BRANDON, S.G.F.: Time and the Destiny of Man. *The Voices of Time,* J.T. Fraser, ed.

BREMER, STUART A.: *Simulated Worlds: A Computer Model of National Decision-Making.* Princeton: Princeton Univ. Press, 1976

BREMMERMAN, HANS J.: Complexity and Transcomputability. 168-174 of Duncan and Weston-Smith

BROOKS, F.P. JR.: *The Mythical Man-Month: Essays on software engineering.* Reading, MA: Addison-Wesley, 1975

BURKE, T.E.: *The Philosophy of Popper.* Manchester Univ. Press, 1983

CAMPBELL, D.M. and HIGGINS, J.C., eds.: *Mathematics: People, Problems, Results.* 3 vols. Belmont: Wordsworth International, 1984

CAPEK, M.: *The Concepts of Space and Time.* Dordrecht/Boston: D. Reidel Publishing, 1976

CAPONEGRI, A.R.: *Time and Idea.* Chicago: Henry Regnery, 1953

CARBONELL, JAIME G.: *Metaphor Comprehension.* Computer Science Dept., Carnegie-Mellon University, May 1981

CARD, S.K., MORAN. T.P. and NEWELL, ALLEN: *The Psychology of Human-Computer Interaction.* Hillsdale, N.J.: Lawrence Erlbaum Associates, 1984

CHAITIN, GREGORY J.: *Randomness and Mathematical Proof.* Scientific American, 1975

CLARK, SIR KENNETH: *What is a Masterpiece?* London: Thames and Hudson, 1979

CLARK, SIR KENNETH: *The Art of Humanism.* Chap. 2. New York: Harper & Row, 1983

CLEMENT, JOHN, LOCHHEAD, JACK and SOLOWAY, ELLIOT: *Translating between Symbol Systems: Isolating a Common Difficulty in Solving Algebra Word Problems.* Cognitive Development Project, Dept. of Physics and Astronomy, University of Massachusetts, Amherst, 1979

CORNFORD, F.M.: The Invention of Space. Reprinted in *The Concepts of Space and Time,* by Milic Capek. Dordrecht/Boston: D. Reidel Publishing, 1976

COSRIMS: *The Mathematical Sciences: A Collection of Essays.* Edited by the National Research Council's Committee on the Support of Research in the Mathematical Sciences. Cambridge: MIT Press, 1969

CRICK, FRANCIS: *Of Molecules and Men.* Seattle: University of Washington Press, 1966

CROMBIE, A.C., ed.: *Scientific Change.* London: Heinemann, 1963

DAHLQUIST, G. and BJORCK, A.: *Numerical Methods.* Englewood Cliffs, N.J.: Prentice-Hall, 1974

DASTON, LORRAINE J.: *The Reasonable Calculus: Classical Probability Theory, 1650-1840.* Ph.D. dissertation, Harvard University, Cambridge, 1979

DASTON, LORRAINE J.: Probabilistic Expectation and Rationality in Classical Probability Theory. *Historia Mathematics,* 7:234-260, 1980

DAVID, F.N.: *Games, Gods and Gambling.* New York: Hafner, 1962

DAVIS, ERNEST S.: Limits and Inadequacies in Artificial Intelligence, *No Way,* P.J. Davis and D. Park, eds., New York: W.H. Freeman, 1986

DAVIS, ERNEST S.: Reasoning, Common Sense. *Encyclopaedia of Artificial Intelligence.* New York: Wiley, 1986

DAVIS, MARTIN: *Computability and Unsolvability.* New York: McGraw-Hill, 1958

DAVIS, MARTIN, ed.: *The Undecidable.* New York: Raven Press, 1965

DAVIS, MARTIN: What is a Computation? 241-267 of L.A. Steen, *Mathematics Today*

DAVIS, P.J.: *Numerical Analysis: COSRIMS: The Mathematical Sciences,* 128-137. Cambridge: MIT Press, 1969

DAVIS, P.J.: "Fidelity in Mathematical Discourse: Is one and one really two?" *American Math. Monthly,* 79:252-263, 1972

DAVIS, P.J.: When Mathematics Says No: Impossibilities in the Mathematical Field. *No Way,* P.J. Davis and D. Park, eds. New York: W.H. Freeman, 1986

DAVIS, P.J. and HERSH, REUBEN: *The Mathematical Experience.* Boston: Birkhäuser, 1981

DAVIS, P.J. and PARK, D., eds: *No Way: The Nature of the Impossible,* New York: W.H. Freeman, 1986

DAVIS, RANDALL and LEVAT, DOUGLAS B.: *Knowledge-Based Systems in Artificial Intelligence.* New York: McGraw-Hill, 1982

DEKEN, JOSEPH G.: *The Electronic Cottage.* New York: Bantam, 1982

DEKEN, JOSEPH G.: *Computer Images.* New York: Stewart, Tabori and Chang, 1983

DE BRUIJN, N.G.: "A Survey of the Project Automath". *Essays on Combinatory Logic, Lambda Calculus, and Formalism.* J.P. Seldin and J.R. Hindley, eds. New York: Academic Press, 1980

DE LA MARE, WALTER: *Behold, This Dreamer.* London: Faber, 1939

DE MILLO, R.A., LIPTON, R.J., and PERLIS, A.J.: Social Processes and Proofs of Theorems and Programs, *Communications of the ACM.* 22:271-280, 1979. Reprinted in *Mathematics,* Campbell and Higgins, eds.

DENNETT, DANIEL C.: The Role of the Computer Metaphor in Understanding the Mind. *Computer Culture,* H.R. Pagels, ed.

DERTOUZOS, MICHAEL L. and MOSES, JOEL, eds.: *The Computer Age: A Twenty-Year View.* Cambridge: MIT Press, 1979

DIJKSTRA, E.W.: *A Discipline of Programming.* Englewood Cliffs, N.J.: Prentice-Hall, 1976

DIJKSTRA, E.W., DE MILLO, R.A., LIPTON, R.J. and PERLIS, A.J.: A polemical exchange between Dijkstra and the others. *ACM SIGSOFT, Software Engineering Notes.* 3(2):14-17 April 1978

DREYFUS, HUBERT L.: *What Computers Can't Do.* Revised Edition. New York: Harper & Row, 1979

DREYFUS, HUBERT L.: A Framework for Misrepresenting Knowledge. 124-136 of *Philosophical Perspectives in Artificial Intelligence*, M. Ringle, ed. Brighton: Harvester Press, 1979

DUNCAN, RONALD and WESTON-SMITH, MIRANDA, eds.: *The Encyclopedia of Ignorance.* New York: Pergamon Press, 1977; Pocket Books, 1978

ECCLES, SIR JOHN and ROBINSON, D.N.: *The Wonder of Being Human.* New York: Free Press, 1984

EISEMAN, PETER R.: Geometric Methods in Computational Field Dynamics. *ICASE Report 80-11*, April 1970, Langley Field, Va.

EITZEN, S.D.: *Sport in Contemporary Society.* New York: St. Martin's Press, 1979

ELLUL, JACQUES: *L'Empire du Non-Sens.* Paris: Presses Universitaires de France, 1980

D'ESPAGNAT, BERNARD: *In Search of Reality.* New York: Springer Verlag, 1983

EVANS, CHRISTOPHER: *The Micro Millennium.* New York: Washington Square Press, 1979

FEIGENBAUM, E.A. and MCCORDUCK, P.: *The Fifth Generation.* Reading, MA: Addison-Wesley, 1983

FINE, TERRENCE L.: *Theories of Probability: An Examination of Foundations.* New York: Academic Press, 1973

FINKELSTEIN, M.O.: *Quantitative Methods in Law.* New York: Free Press-Macmillan, 1978

FOLEY, JAMES D. and VAN DAM, ANDRIES: *Fundamentals of Interactive Computer Graphics.* Reading, MA: Addison-Wesley, 1982

FORD, JOSEPH: *How Random is a Coin Toss?* School of Physics, Georgia Institute of Technology, 1983

FORRESTER, JAY W.: *Counterintuitive Behavior of Social Systems.* Technology Review 73, January 1971

VON FRANZ, M.L., ed: Time and Syncronicity in Analytical Psychology. *The Voices of Time*, J.T. Fraser, ed

FRASER, J.T., ed.: *The Voices of Time.* Amherst: Univ. of Massachusetts Press, 1981

GAFFNEY, MATTHEW P. and STEEN, LYNN ARTHUR: *Annotated Bibliography of Expository Writing in the Mathematical Sciences.* Mathematical Association of America, 1976

GILLISPIE, C.C.: Intellectual Factors in the Background of Analysis by Probability, *Scientific Change*, A.C. Crombie, ed., 431-453

GILLISPIE, C.C.: *The Edge of Objectivity.* Princeton: Princeton Univ. Press, 1970

GODWIN, JOHN: *The Mating Trade.* New York: Doubleday, 1973

GOLDSTINE, HERMAN H.: *The Computer from Pascal to von Neumann.* Princeton: Princeton Univ. Press, 1972

GOLDSTINE, HERMAN H.: *A History of Numerical Analysis from the 16th through the 19th Century.* Heidelberg: Springer-Verlag, 1977

GOOD, I.J.: *Speculation Concerning the First Ultra-Intelligent Machine.* Princeton: Institute for Defense Analyses, 1964

GOOD, I.J.: Subject Index of Speculation Concerning the First Ultraintelligent Machine. *Advances in Computers*, 6:31-88, 1965

GOOD, I.J.: Dynamic Probability, Computer Chess, and the Measurement of Knowledge. *Machine Intelligence*, 8:139-150, 1977

GRABINER, JUDITH V.: Is Mathematical Truth Time-Dependent? *Amer. Math. Monthly*, April 1974, 354-365

GREGORY, RICHARD L.: Consciousness. 274-281 of Duncan and Weston-Smith

Gregory, R.L.: *The Intelligent Eye.* London: Weidenfeld and Nicolson, 1970

Grunbaum, Branko: Shouldn't We Teach GEOMETRY? *Two Year College Mathematics Journal,* 12:232-238, 1981

Hacking, Ian: *The Emergence of Probability.* Cambridge: Cambridge University Press, 1975

Hamming, R.W.: *Numerical Methods for Scientists and Engineers.* 2d edition. New York: McGraw-Hill, 1973

Hirt, C.W. and Ramshaw, J.D.: Prospects for Numerical Simulation of Bluff-Body Aerodynamics. *Aerodynamic Drag Mechanisms of Bluff Bodies and Road Vehicles,* Sovran, Morel, and Mason, eds., New York: Plenum, 1978

Hofstadter, Douglas: *Godel Escher Bach: the Eternal Golden Braid.* New York: Harper & Row, 1979

Hofstadter, Douglas R. and Dennett, Daniel C.: *The Myth of the Computer. The Mind's I: Fantasies and Reflections on Self and Soul.* New York: Basic Books, 1983

Hoppenstadt, F.C.: *Mathematical Methods of Population Biology.* Cambridge: Cambridge Univ. Press, 1982

Houts, Paul L., ed.: *The Myth of Measurability.* New York: Hart Publishing Co., 1977

Huber, R.K., ed.: *Systems Analysis and Modeling in Defense.* New York: Plenum, 1984

Huxley, Aldous: *Literature and Science.* New York: Harper & Row, 1963

Iano, Richard: *Is Education a Science? No Way!* In P.J. Davis & D. Park.

Jaki, Stanley L.: *Brain, Mind and Computers.* South Bend, Indiana: Gateway Editions, 1969

Johnson, Deborah G.: *Computer Ethics,* Englewood Cliffs: Prentice-Hall, 1985

Kemeny, J.G.: The Case for Computer Literacy. *Daedalus,* Spring 1983, 211-230

Kent, E.W.: *The Brains of Men and Machines.* New York: McGraw-Hill, 1981

Kerr, Donald M., Braithwaite, Karl, Metropolis, N., Sharp, David H., and Rota, Gian-Carlo, eds.: *Science, Computers, and the Information Onslaught—A collection of Essays.* Orlando: Academic Press, 1984

Knuth, Donald E.: *The Art of Computer Programming. Volume I. Fundamental Algorithms.* Reading, MA: Addison-Wesley, 1968; Volume II 1969, Volume III, 1973

Knuth, Donald E.: Computer Science and its Relation to Mathematics. *American Math. Monthly,* April 1974

Knuth, Donald E.: Mathematics and Computer Science: Coping with Finiteness. *Science* 194:1235-1242, 1976

Kline, Morris: *Mathematical Thought from Ancient to Modern Times.* Oxford: Oxford Univ. Press, 1972

Koblitz, N.: Mathematics as Propaganda. *Mathematics Tomorrow,* 111-120, Lynn Arthur Steen, ed. New York: Springer-Verlag, 1981

Kohlberg, Lawrence: *The philosophy of Moral Development.* New York: Harper & Row, 1981

Kosslyn, Stephen M. and Hatfield, Gary C.: Representation Without Symbol Systems. To appear in *Social Research,* special issue on Cognitive Science

Koyre, A.: *From the Closed World to the Infinite Universe.* Baltimore: Johns Hopkins Press, 1957

Kuhn, Thomas S.: *The Structure of Scientific Revolutions.* 2d edition. Chicago: Univ. of Chicago Press, 1970

Ladd, John, ed.: *Ethical Relativism.* New York: Wadsworth, 1973

Ladd, John: *Ethics and the Computer Revolution.* Providence: Brown University, Department of Philosophy, 1985

Lakatos, Imre: *Proofs and Refutations.* J. Worral and E. Zahar, eds. Cambridge: Cambridge Univ. Press, 1976

Lakoff, George and Johnson, Mark: *Metaphors We Live By.* Chicago: Univ. of Chicago Press, 1980

Landauer, Rolf: Wanted: A Physically Possible Theory of Physics. *IEEE Spectrum* 4(4):105-109, 1967

Landauer, Rolf: Fundamental Physical Limitations of the Computational Process. pp. 161-170 of Pagel, H.R., ed.

Langer, Susanne K.: *Mind: An Essay on Human Feeling.* Baltimore: The Johns Hopkins Univ. Press, 1982

Lax, Peter D., chairman: Report of the Panel on Large-Scale Computing in Science and Engineering. *National Science Report 83-13.* Washington, D.C., 1983

Leavis, F.R.: *The Living Principle.* Oxford: Oxford Univ. Press, 1975

Lebowitz, Michael: *The Nature of Generalization in Understanding.* Dept. of Computer Science, Columbia University, 348-353

Lem, Stanislaw: *The Cyberiad.* New York: Avon, 1976

Lem, Stanislaw: *Perfect Vacuum.* New York: Harcourt Brace Jovanovich, 1979

Leplin, Jarrett, ed: *Scientific Realism.* Berkeley: Univ. of California Press, 1984

Lineberry, W.P., ed: *The Business of Sports (The Reference Shelf Series).* New York: H.W. Wilson Company, 1973

Locke, John: *An Essay Concerning Human Understanding.* Abridged Version edited by Maurice Cranston. New York: Collier-Macmillan, 1965. Especially "Of the Reality of Knowledge"

MacCormack, Earl R.: *Metaphor and Myth in Science and Religion.* Durham, NC: Duke Univ. Press, 1976

Malina, F.J., ed: *Visual Art, Mathematics and Computers.* Oxford: Pergamon Press, 1977

Mandelbrot, Benoit B.: *Fractals: Form, Chance and Dimension.* San Francisco: W.H. Freeman & Co., 1977

Maritain, Jacques: *The Dream of Descartes.* New York: Philosophical Library, 1944

Mayer, W.J.: *Concepts of Mathematical Modeling.* New York: McGraw-Hill, 1984

McCloskey, Donald N.: The Rhetoric of Economics. *Journal of Economic Literature,* 1-70, 1983

McCorduck, Pamela: *Machines Who Think.* San Francisco: W.H. Freeman & Co., 1979

McCorduck, Pamela: Knowledge Technology: The Promise. *Computer Culture,* H.R. Pagels, ed.

Medawar, Sir Peter B.: *The Limits of Science.* New York: Harper & Row, 1984

Metropolis, N., Howlett, J., and Rota, G.-C., eds.: *A History of Computing in the 20th Century.* New York: Academic Press, 1980

Miller, Lance A.: *Natural Language Programming: Styles, Strategies, and Contrasts.* Research Report, Computer Sciences Dept., IBM Research Center, Yorktown Heights, NY

Minsky, Marvin: A Framework for Representing Knowledge. *The Psychology of Computer Vision,* P.H. Winston ed. New York: McGraw-Hill, 1975

Minsky, Marvin L.: Computer Science and the Representation of Knowledge. Dertouzos and Moses, 392-421

MUMFORD, LEWIS: *The Myth of the Machine: The Pentagon of Power.* New York: Harcourt Brace Jovanovich, 1970

MYER, LEONARD B.: *Music, the Arts, and Ideas: Patterns and Predictions in Twentieth-Century Cultures.* Chicago: Univ. Chicago Press, 1967

NEEDHAM, JOSEPH: Poverties and Triumphs of the Chinese Scientific Tradition. Scientific Change, A.C. Crombie, ed., 117-153

NEEDHAM, JOSEPH, ed.: *Science, Religion and Reality,* Reprinted, G. Braziller, 1955

NEWELL, ALLEN: *The Knowledge Level.* Presidential Address, American Association for Artificial Intelligence, AAA 180, Stanford Univ., 1980

NEWELL, ALLEN: *The Heuristic of George Pòlya and its Relation to Artificial Intelligence.* Dept. of Computer Science, Carnegie-Mellon University, 1980

NEWMAN, J.R., ed.: *The World of Mathematics.* Four volumes. New York: Simon & Schuster, 1956

NOLL, A.M.: *Computers and the Visual Arts: A Retrospective View.* SIGGRAPH '82 Art Show, Association for Computing Machinery, 1982

NORDON, DIDIER: *Les Mathématiques Pures N'Existent Pas!* Paris: Actes Sud, 1981

OETTINGER, ANTHONY G.: *Run, Computer, Run: The Mythology of Educational Innovation.* Cambridge: Harvard Univ. Press, 1969

OWEN, D.B., ed.: *On the History of Statistics and Probability.* New York: Marcel Dekker, Inc. 1976

PAGELS, HEINZ R., ed.: *Computer Culture, Annals of the New York Academy of Science,* Vol. 426, 1984

PAGELS, HEINZ R., et al: Panel Discussion: Has Artificial Intelligence Research Illuminated Human Thinking? *Computer Culture,* H.R. Pagels, ed.

PAPERT, SEYMOUR: *Mindstorms.* New York: Basic Books, 1980

PARK, DAVID: *The Image of Eternity: Roots of Time in the Physical World.* Amherst, MA: Univ. of Massachusetts Press, 1980

PARK, DAVID: "What Do You Think About When You Think About Time?" *Berkshire Review,* 20:24-34, 1985

PARLETT, B.: Progress in Numerical Analysis. *SIAM Review* 20:443-456, 1978

PEDOE, DAN: *Geometry and the Visual Arts.* New York: St. Martin's Press, 1976; Dover, 1983

PIERCE, JOHN R.: *The Science of Musical Sound.* New York: Scientific American Library, W.H. Freeman, 1983

POLLACK, SEYMOUR V., ed.: Studies in Computer Science. *Studies on Mathematics* 22 Mathematical Association of America

POPPER, KARL R.: *The Open Society and Its Enemies.* London: Routledge and Kegan Paul, 13th impression, 1980

POPPER, KARL R. and ECCLES, JOHN C.: *The Self and Its Brain.* New York: Springer-Verlag, 1977

PURCELL, EDWARD A.: *The Crisis of Democratic Theory; Scientific Naturalism and the Problem of Value.* Lexington: Univ. of Kentucky Press, 1973

PLYSHYN, ZENON W., ed.: *Perspectives on the Computer Revolution.* Englewood Cliffs, NJ: Prentice-Hall, 1970

RABIN, M.O.: Theoretical Impediments to Artificial Intelligence. *Information Processing* 74:615-619, 1974

RAPHAEL, BERTRAM: *The Thinking Computer.* San Francisco: W.H. Freeman, 1976

RAPP, F.: *Analytical Philosophy of Technology*. Dordrecht: D. Reidel, 1981

READ, HERBERT: *Art and Society*. New York: Macmillan, 1937

RICHARDS, J.: "The Reception of a Mathematical Theory: Non Euclidean Geometry in England 1868-1883." Barry Barnes and Stephen Shapin, eds. *Natural Order: Historical Studies of Scientific Culture*, Sage Publications, Beverly Hills, 1979

RINGLE, MARTIN, ed.: *Philosophical Perspectives on Artificial Intelligence*. Brighton: Harvester Press, 1979

RINGLE, MARTIN: *Philosophy and Artificial Intelligence*. 1-20 of M. Ringle, ed.

RISSLAND, EDWINA L.: *Understanding Understanding Mathematics*. Cognitive Science 2, 1978

RUSSELL, BERTRAND: *A History of Western Philosophy and its Connection with Political and Social Circumstances from the Earliest Times to the Present Day*. New York: Simon & Schuster, 1945

SACKMAN, HAROLD: *Computers, System Science and Evolving Society*. New York: John Wiley, 1967

SALTON, G.: What is Computer Science? *Journal of the Association for Computing Machinery* 19:1-2, 1972

SCHANK, ROGER C. and COLBY, KENNETH M.: *Computer Models of Thought and Language*. San Francisco: W.H. Freeman, 1973

SCHWARTZ, J.T.: The Pernicious Influence of Mathematics on Science. *Proceedings of the 1960 International Congress on Logic, Methodology and Philosophy of Science*. Stanford: Stanford Univ. Press, 1962

SCHWARTZ, J.T.: The Limits of Artificial Intelligence. *Encyclopedia of Artificial Intelligence*. New York: John Wiley, 1986

SEARLE, JOHN R.: Minds, Brains and Programs. *The Behavioral and Brain Sciences* 3:417-457, 1980. Also in: John Haugeland ed., "*Mind Design: Philosophy, Psychology, Artificial Intelligence*", M.I.T. Press, 1981

SEARLE, JOHN R.: *Minds, Brains and Science*. Cambridge: Harvard Univ. Press, 1984

SHEROVER, C.M., ed.: *The Human Experience of Time*. New York: New York Univ. Press, 1975

SIGGRAPH '82 Art Show: *An Exhibition Highlighting the Recent Achievements of Artists Working with Computers*, Association for Computing Machinery, 1982

SIMON, HERBERT A.: Artificial Intelligence Research Strategies in the Light of AI Models of Scientific Discovery. *Naval Research Reviews*, ONR, XXXIV: 1-16, 1982

SIMONS, GEOFF: *Are Computers Alive?* Boston: Birkhäuser, 1983

SIWOFF, S., ed.: *The Book of Baseball Records*. New York: Seymour Siwoff, 1980

SLOMAN, AARON and CROUCHER, MONICA: *Why Robots Will Have Emotions*. Cognitive Studies Programme, University of Sussex, England, 197-212

STEEN, LYNN ARTHUR, ed.: *Mathematics Today: Twelve Informal Essays*. New York: Springer-Verlag, 1978

STEINER, GEORGE: *Bluebeard's Castle: Some notes towards the redefinition of culture*. New Haven: Yale University Press, 1971

STOCKMEYER, L.J. and CHANDRA, A.K.: Intrinsically Difficult Problems. *Scientific American*, January 1978

TOTH, I.: "Gott und Geometrie." *Evolutionstheorie und ihre Evolution*. Vortragsreihe der Universitat Regensburg zum 100. Todestag von Charles Darwin, 141-204

TOWNEND, M.S.: *Mathematics in Sport*. New York: John Wiley & Sons, 1984

TRAUB, J.F.: *Coping with Complexity.* Department of Computer Science, Columbia University, 1983

TRAUB, J.F., ed.: *Cohabiting with Computers.* Los Altos: William Kaufmann Inc., 1985

TRUESDELL, CLIFFORD A.: The Computer: Ruin of Science and Threat to Mankind. *An Idiot's Fugitive Essays.* New York: Springer-Verlag, 1984

TURKLE, SHERRY: *The Second Self: Computers and the Human Spirit.* New York: Simon & Schuster, 1984

TYMOCZKO, THOMAS: *Making Room for Mathematicians in the Philosophy of Mathematics.* Smith College, June 1981

VALERY, PAUL: *Occasions.* Princeton: Princeton Univ. Press, 1970

VROOMAN, J.R.: *René Descartes.* New York: G.P. Putnam's Sons, 1970

WEBER, MAX: *The Protestant Ethic and the Spirit of Capitalism. 1904-1905.* Translated by Talcott Parsons, 1930. Foreword by R.H. Tawney. New York: Scribner, 1958

WEINBERG, STEVEN: "*The First Three Minutes: A modern view of the origin of the Universe*". Basic Books, 1977

WEISSKOPF, VICTOR F.: "The Origin of the Universe". *American Scientist* September-October, 473-480, 1983

WEIZENBAUM, JOSEPH: "*Computer Power and Human Reason: From Judgement to Calculation*", W.H. Freeman, 1976

WEIZENBAUM, JOSEPH: Once More: The Computer Revolution. 439-458 of Dertouzos and Moses, eds.

WHITROW, G.J.: *The Natural Philosophy of Time.* First edition: Nelson, 1961; second edition: Oxford: Oxford Univ. Press, 1980

WIGNER, EUGENE P.: The Unreasonable Effectiveness of Mathematics in the Natural Sciences. *Comm. in Pure and Applied Mathematics,* 13:1-14, 1960. Reprinted in Vol. III of Campbell and Higgins

WILDER, RAYMOND L.: *Evolution of Mathematical Concepts: An Elementary Study.* New York: John Wiley & Sons, 1968

WINSTON, PATRICK HENRY and BROWN, RICHARD HENRY: *AI: an MIT Perspective.* Cambridge: MIT Press, 1979

WOLFSON, HARRY A.: *The Philosophy of the Kalam.* Cambridge: Harvard Univ. Press, 1976

WOLFSON, HARRY A.: *Repercussions of the Kalam in Jewish Philosophy.* Cambridge: Harvard Univ. Press, 1979

WOS, LARRY, OVERBEEK, ROSS, LUSK, EWING, and BOYLE, JIM: *Automated Reasoning.* Englewood Cliffs, New Jersey: Prentice-Hall, 1984

YOLTON, JOHN W.: *Appearance and Reality. Dictionary of the History of Ideas, vol I.* New York: Scribners, 1968

YOUNGBLOOD, GENE: *Towards Autonomous Reality Communities.* SIGGRAPH '82 Art Show, Association for Computing Machinery, 1982

ZIMAN, J.: *Reliable Knowledge.* Cambridge: Cambridge Univ. Press, 1978

ZIMAN, J.: *Puzzles, Problems and Enigmas.* Cambridge: Cambridge Univ. Press, 1981

Index

Index